Electric Motors and Drives

Fundamentals, types and applications

Second edition

Austin Hughes

*Senior Lecturer, Department of Electronic and Electrical
Engineering, University of Leeds*

NEWNES

Newnes
An imprint of Butterworth-Heinemann Ltd
Linacre House, Jordan Hill, Oxford OX2 8DP

 A member of the Reed Elsevier plc group

OXFORD LONDON BOSTON
MUNICH NEW DELHI SINGAPORE SYDNEY
TOKYO TORONTO WELLINGTON

First published 1990
Reprinted 1990, 1991, 1992

Second edition 1993
Reprinted 1994

British Library Cataloguing in Publication Data
Hughes, Austin
    Electric motors and drives – 2nd ed.
    1. Electric motors
    I. Title
    621.46'2

ISBN 0 7506 1741 1

Typeset by Vision Typesetting, Manchester
Printed and bound in Great Britain by
Biddles Ltd, Guildford and King's Lynn

CONTENTS

9 SYNCHRONOUS, SWITCHED RELUCTANCE AND BRUSHLESS D.C. DRIVES

PREFACE

Like its predecessor, the second edition of this book is intended primarily for non-specialist users of electric motors and drives. My original aim was to bridge the gap between specialist textbooks (which are pitched at a level which is too academic for the average user) and the more prosaic 'handbooks' which are full of useful detail but provide little opportunity for the development of any real insight or understanding. It has therefore been a pleasure to receive so much agreeable feedback from readers in industry, and to learn that the level of treatment has met with widespread approval. Having deliberately set out with the specific intention of not writing yet another student textbook, I nevertheless had a sneaking suspicion that it might also appeal to undergraduates, so it has been a bonus to learn from students in Leeds and elsewhere that they welcome the book as a gentle introduction to the subject.

The second edition has been revised to take account of the many constructive suggestions received from readers, and to expand the sections dealing with current developments in the motors and drives scene. Amongst the topics which are given more exposure are vector (field-oriented) control of induction motor drives, which is now available as an off-the-shelf item; switched reluctance drives which are rapidly expanding into new applications; and power switching devices, where devices such as the IGBT have recently advanced into the inverter market. The cycloconverter drive was a notable omission from the first edition, so a new section has been added in Chapter 7.

Most of the changes to the text are intended to clarify understanding, and I hope that the addition of a section of questions and answers at the end of each chapter will help in this respect.

Only older readers will appreciate how radical and exciting the changes have been over the past twenty or so years, so perhaps a few words are appropriate to put the current scene into perspective. For more than a century, many different types of motor were developed, and each became closely associated with a particular application. Traction, for example, was seen as the exclusive preserve of the series d.c. motor, whereas the shunt d.c. motor, though outwardly indistinguishable, was seen as being quite unsuited to traction applications. The cage induction motor was (and still is) the most numerous type but was judged as being suited only to applications which called for constant speeds.

The reason for the plethora of motor types was that there was no easy way of varying the supply voltage and/or frequency to obtain speed control, and designers were therefore forced to seek ways of providing for control of speed within the motor itself. All sorts of ingenious arrangements and interconnections of the various motor windings were invented, but even the best motors had a limited range of operating characteristics, and all of them required bulky control gear which was manually or electromechanically operated, making it difficult to arrange automatic or remote control.

All this changed from the early 1960s, when power electronics began to make a real impression. The first breakthrough came with the thyristor, which provided a relatively cheap, compact, and easily controlled variable-voltage supply for the d.c. motor. Some minor changes were called for in the motor to accommodate the less than perfect 'd.c.' which the thyristor rectifier provided, but the flexibility offered by the electronic control of speed and torque meant that a truly adaptable workhorse was at last a practicable proposition. In the 1970s the second major breakthrough resulted from the development of variable-frequency inverters, capable of providing a three-phase supply suitable for induction motors. The induction motor, which for so long had been restricted to constant speed applications, was at last able to compete in the controlled-speed arena.

These major developments resulted in the demise of many of the special motors, leaving the majority of applications in the hands of comparatively few types, and the emphasis has now shifted from complexity inside the motor to sophistication in supply and control arrangements. From the user's point of view this is a mixed blessing. Greater flexibility and superior levels of performance are available, and there are fewer motor types to consider. But if anything more than constant speed is called for, the user will be faced with the purchase of a complete drive system, consisting of a motor together with its associated power electronics package. To choose wisely requires not only some knowledge of motors, but also the associated power electronics and the control options which are normally provided.

An awareness of the user's potential difficulties has been uppermost in shaping the contents. The aim throughout is to provide the reader with an understanding of how each motor and drive system works, in the belief that it is only by knowing what should happen that informed judgements and sound comparisons can be made. Given that the book is aimed at non-specialists from a range of disciplines, introductory material on motors and power electronics is clearly necessary, and this is presented in the first two chapters. Many of these basic ideas crop up frequently throughout the book, so unless the reader is already well-versed in the fundamentals it would be wise to absorb the first two chapters before tackling the later material.

The remainder of the book explores most of the widely-used modern types of motor and drive, including conventional and brushless d.c., induction motors (mains and inverter-fed), stepping motors, synchronous motors (mains and converter-fed) and reluctance motors. The d.c. motor drive and the induction motor drive are given most weight, reflecting their dominant position in terms of numbers. Understanding the d.c. drive is particularly important because it is so widely used as the yardstick by which other drives are measured. Users who develop a good grasp of the d.c. drive will find their know-how invaluable in dealing with all other types, particularly if they can establish a firm grip on the philosophy of the control scheme.

Applications are spread throughout the text in order to emphasise the fact that there is no longer any automatic correlation between motor type and particular application. Similarities between the various motors and drives have also been given emphasis in order to underline the fact that apparently different types have a great deal in common at the fundamental level. Recognising this degree of commonality is important from the user's viewpoint, but is seldom given weight (albeit for obvious reasons) in sales literature.

The style of the book reflects my own preference for an informal approach, in which the difficulty of coming to grips with new ideas is not disguised. Deciding on the level at which to pitch the material was originally a headache, but experience suggested that a mainly descriptive approach with physical explanations would be most appropriate, with mathematics kept to a minimum to assist digestion. Feedback from the first edition has encouraged me to maintain this approach.

1

ELECTRIC MOTORS

INTRODUCTION

Electric motors are so much a part of everyday life that we seldom give them a second thought. When we switch on an electric drill, for example, we expect it to run rapidly up to the correct speed, and we don't question how it knows what speed to run at, nor how it is that once enough energy has been drawn from the supply to bring it up to speed, the power drawn falls to a very low level. When we put the drill to work it draws more power, and when we finish the power drawn from the mains reduces automatically, without intervention on our part.

The humble motor, consisting of nothing more than an arrangement of copper coils and steel laminations, is clearly rather a clever energy converter, which warrants serious consideration. By gaining a basic understanding of how the motor works, we will be able to appreciate its potential and its limitations, and (in later chapters) see how its already remarkable performance can be even further improved by the addition of external controls.

This chapter deals with the basic mechanisms of motor operation, so readers who are already familiar with such matters as magnetic flux, magnetic and electric circuits, torque, and motional e.m.f can probably afford to skip most of it. In the course of the discussion, however, several very important general principles and guidelines emerge. These apply to all types of motor and are summarised in the concluding section. Experience shows that anyone who has a good grasp of these basic principles will be well equipped to weigh the pros and cons of the different types of motor, so all readers are urged to absorb them before tackling other parts of the book.

PRODUCING ROTATION

Nearly all motors exploit the force which is exerted on a current-carrying conductor placed in a magnetic field. The force can be demonstrated by placing a bar magnet near a wire carrying current, but anyone who tries the experiment will probably be disappointed to discover how feeble the force is, and will doubtless be left wondering how such an unpromising effect can be used to make effective motors.

We will see that in order to make the most of the mechanism, we need to arrange for there to be a very strong magnetic field, and for it to interact with many conductors, each carrying as much current as possible. We will also see later that although the magnetic field (or 'excitation') is essential to the working of the motor, it acts only as a catalyst, and all of the mechanical output power comes from the electrical supply to the conductors on which the force is developed. It will emerge later that in some motors the parts of the machine responsible for the excitation and for the energy-converting functions are separate and easily distinguished. In the d.c. motor, for example, the excitation is provided either by permanent magnets or by field coils wrapped around clearly-defined projecting field poles on the stationary part, while the conductors on which force is developed are on the rotor and supplied with current via sliding brushes. In many motors, however, there is no such clearcut physical distinction between the 'excitation' and the 'energy-converting' parts of the machine, and a single stator winding serves both purposes. Nevertheless, we will find that identifying and separating the excitation and energy-converting functions is always helpful to understanding how motors of all types operate.

Returning to the matter of force on a single conductor, we

will look first at what determines the magnitude and direction of the force, before turning to ways in which the mechanism is exploited to produce rotation. The concept of the magnetic circuit will have to be explored, since this is central to understanding why motors have the shapes they do. Magnetic flux and flux density will crop up continuously in the discussion, so a brief introduction to the terms is included for those who are not already familiar with the ideas involved.

Electromagnetic force

When a current-carrying conductor is placed in a magnetic field, it experiences a force. Experiment shows that the magnitude of the force depends directly on the current in the wire, and the strength of the magnetic field, and that the force is greatest when the magnetic field is perpendicular to the conductor. The direction of the force is shown in Figure 1.1: it is at right angles to both the current and the magnetic field. With the magnetic field vertically downwards, and the current flowing into the paper, the force is horizontal and to the left. If either the field or the current is reversed, the force acts to the right, and if both are reversed, the force will remain to the left.

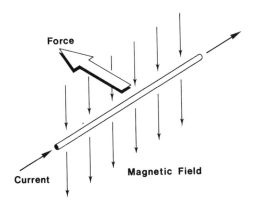

Figure 1.1 *Electromagnetic force on current-carrying conductor in a magnetic field*

These simple relationships have a pleasing feel to them: if we double either the current or the field strength, we double the force, while doubling both causes the force to increase by a factor of four. But what about quantifying the force? We need to express the force in terms of the product of the current and the magnetic field strength, and this turns out to be very straightforward when we describe the magnetic field in terms of magnetic flux density, B.

Magnetic flux and flux density

The familiar patterns made by iron filings in the vicinity of a bar magnet offer us clear guidance in our quest for an effective way of picturing and quantifying the magnetic field. The tendency of the filings to form themselves into elegant curved lines, as shown in Figure 1.2, is indicative of the presence of a magnetic field; and their orientation immediately suggests the notion of a direction of the field at each point.

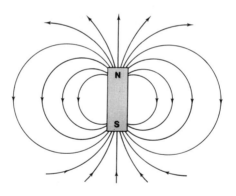

Figure 1.2   *Magnetic flux lines surrounding a bar magnet*

Experiment shows that where the field is strongest (at the ends of the magnet), the lines are close together, while in the weaker field regions (remote from the ends of the magnet), the lines are far apart. In particular, we find that if we focus on a particular pair of adjacent lines, the strength of the field halves when the distance between the lines doubles.

If we couple these observations to the idea that between each pair of lines (and assuming a fixed distance into the paper) there is a fixed quantity of magnetic flux, we are led to the concept of magnetic flux density. When the lines are close together, the 'tube' of flux is squashed into a smaller space, giving a high flux density; whereas when the lines spread further apart the same tube of flux has more breathing space and the flux density is lower. In each case the direction of the flux density is indicated by the prevailing direction of the lines.

All that remains is to specify units for quantity of flux, and flux density. In the SI system, the unit of magnetic flux is the weber (Wb). If one weber of flux is distributed uniformly across an area of one square metre perpendicular to the flux, the flux density is one weber per square metre, or one tesla (T). In general if a flux Φ is distributed uniformly across an area A, the flux density is given by

$$B = \frac{\Phi}{A} \qquad (1.1)$$

In the motor world we are unlikely to encounter more than a few milliwebers of flux, and a small bar magnet would probably only produce a few microwebers. On the other hand, values of flux density are typically around 1 tesla in most motors, which is a reflection of the fact that although the quantity of flux is small, it is also spread over a small area.

Force on a conductor

We can now quantify the force, and begin to establish some feel for the magnitudes of forces we can exploit in a motor.

The force on a wire of length l, carrying a current I and exposed to a uniform magnetic flux density B is given by the simple expression

$$F = BIl \qquad (1.2)$$

In equation 1.2, F is in newtons when B is in tesla, I in amps, and l in metres. (It may come as a surprise that there are no constants of proportionality involved in equation 1.2: this is not a coincidence, but arises because the unit of current (the ampere) is actually defined in terms of force.)

Strictly, equation 1.2 only applies when the current is perpendicular to the field. If this condition is not met, the force on the conductor will be less; and in the extreme case where the current was in the same direction as the field, the force would fall to zero. However, every sensible motor designer knows that to get the best out of the magnetic field it has to be perpendicular to the conductors, and so it is safe to assume in the subsequent discussion that B and I are always perpendicular.

The reason for the very low force detected in the experiment with the bar magnet is revealed by equation 1.2. To obtain a high force, we must have a high flux density, and a lot of current. The flux density at the ends of a bar magnet is low, perhaps 0.1 tesla, so a wire carrying 1 amp will experience a force of only 0.1 N (approximately 100 gm wt) per metre. Since the flux density will be confined to perhaps 1 cm across the end face of the magnet, the total force on the wire will be only 1 gm. This would be barely detectable, and is too low to be of any use in a decent motor. So how is more force obtained?

The first step is to obtain the highest possible flux density. This is achieved by designing a 'good' magnetic circuit, and is discussed next. Secondly, as many conductors as possible must be packed in the space where the magnetic field exists, and each conductor must carry as much current as it can without heating up to a dangerous temperature. In this way, impressive forces can be obtained from modestly sized devices, as anyone who has tried to stop an electric drill by grasping the chuck will testify.

MAGNETIC CIRCUITS

The concept of the magnetic circuit stems from the fact that magnetic flux lines always form closed contours. This is illustrated in Figure 1.3, which shows the field produced by a circular solenoid. We note that the field pattern is similar to

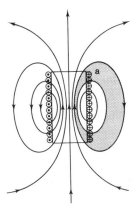

Figure 1.3 *Magnetic field of circular solenoid*

that produced by a bar magnet, and we will see later that in many machines the source of flux can be either a coil (or winding) or a permanent magnet.

Recalling that the space between each pair of lines (and unit depth into the paper) contains a fixed quantity of flux, we can picture one tube of flux as 'starting' at say point **a**, then 'flowing' round a circuit (shown shaded) before arriving back at **a**.

The air surrounding the sources of the field in Figure 1.3 offers a homogeneous path for the flux, so once the tubes of flux escape from the concentrating influence of the source, they are free to spread out into the whole of the surrounding space. The flux density outside the coil is therefore low; and even inside the coil, we would find that the flux densities which we could achieve are still too low to be of use in a motor. What is needed is firstly a way of increasing the flux density, and secondly a means for preventing the flux from spreading out into the surrounding space.

Magnetomotive force (MMF)

One obvious way to increase the flux density is to increase the current in the coil, or to add more turns. We find that if we double the current, or the number of turns, we double the total flux, thereby doubling the flux density everywhere.

We quantify the ability of the coil to produce flux in terms of the magnetomotive force (MMF). The MMF of the coil is simply the product of the number of turns (N) and the current (I), and is thus expressed in ampere-turns. A given MMF can be obtained with a large number of turns of thin wire carrying a low current, or a few turns of thick wire carrying a high current: as long as the product NI is constant, the MMF is the same.

Electric circuit analogy and reluctance

We have seen that the magnetic flux which is set up is proportional to the MMF driving it. This points to a parallel with the electric circuit, where the current (amps) which flows is proportional to the EMF (volts) driving it.

In the electric circuit, current and EMF are related by Ohm's law, which is

$$\text{Current} = \frac{\text{EMF}}{\text{resistance}} \quad \text{i.e.} \quad I = \frac{V}{R} \quad (1.3)$$

For a given source EMF (volts), the current depends on the resistance of the circuit, so to obtain more current, the resistance of the circuit has to be reduced.

We can make use of an equivalent 'magnetic Ohm's law' by introducing the idea of reluctance (Λ). The reluctance gives a measure of how difficult it is for the magnetic flux to complete its circuit, in the same way that resistance indicates how much opposition the current encounters in the electric circuit. The magnetic Ohm's law is then

$$\text{Flux} = \frac{\text{MMF}}{\text{reluctance}} \quad \text{i.e.} \quad \Phi = \frac{NI}{\Lambda} \quad (1.4)$$

We see from equation 1.4 that to increase the flux for a given MMF, we need to reduce the reluctance of the magnetic circuit. In the case of the example (Figure 1.3), this means we must replace as much as possible of the air path (which is a 'poor' magnetic material, and therefore constitutes a high reluctance) with a 'good' magnetic material, thereby reducing the reluctance and resulting in a higher flux for a given MMF.

The material which we choose is good quality magnetic steel, which for historical reasons is usually referred to as 'iron'. This brings two desirable benefits, as shown in Figure 1.4.

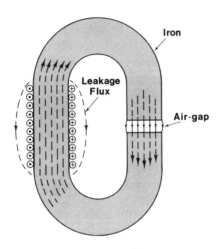

Figure 1.4 *Simple magnetic circuit with air-gap*

Firstly, the reluctance of the iron paths is very much less than the air paths which they have replaced, so the total flux produced for a given MMF is much greater. (Strictly speaking therefore, if the MMFs and cross-sections of the coils in Figures 1.3 and 1.4 are the same, many more flux lines should be shown in Figure 1.4 than in Figure 1.3, but for the sake of clarity similar numbers are indicated.) And secondly, almost all the flux is confined within the iron, rather than spreading out into the surrounding air. We can therefore shape the iron parts of the magnetic circuit as shown in Figure 1.4 in order to guide the flux to wherever it is needed. Note also that inside the iron, the flux density remains uniform over the whole cross-section, there being so little reluctance that there is no noticeable tendency for the flux to crowd to one side or another.

A question which is often asked is whether it is important for the coils to be wound tightly onto the magnetic circuit, and whether, if there is a multi-layer winding, the outer turns are as effective as the inner ones. The answer is that the total MMF is determined solely by the number of turns and the current, and therefore every complete turn makes the same contribution to the total MMF, regardless of whether it happens to be tightly or loosely wound. Of course it does make sense for the coils to be wound as tightly as is practicable, since this not only minimises the resistance of the coil (and thereby reduces the heat loss) but also makes it easier for the heat generated to be conducted away to the frame of the machine.

The air-gap

In motors, we intend to use the high flux density to develop force on current-carrying conductors. We have now seen how to create a high flux density inside the iron parts of a magnetic circuit, but, of course, it is physically impossible to put current-carrying conductors inside the iron. We therefore arrange for an air-gap in the magnetic circuit, as shown in Figure 1.4. We will see shortly that the conductors on which the force is to be produced will be placed in this air-gap region.

If the air-gap is relatively small, as in motors, we find that the flux jumps across the air-gap as shown in Figure 1.4, without any appreciable tendency to balloon out into the surrounding air. With most of the flux lines going straight across the air-gap, the flux density in the gap region has the same high value as it does inside the iron.

In the majority of magnetic circuits consisting of iron parts and one or more air-gaps, the reluctance of the iron parts is very much less than the reluctance of the gaps. At first sight this can seem surprising, since the distance across the gap is so much less than the rest of the path through the iron. The fact that the air-gap dominates the reluctance is simply a reflection of how poor air is as a magnetic medium, compared with iron. To put the comparison in perspective, if we calculate the reluctances of two paths of equal length and cross-sectional area, one being in iron and the other in air, the reluctance of the air path will typically be 1000 times greater than the reluctance of the iron path.

Returning to the analogy with the electric circuit, the role of the iron parts of the magnetic circuit can be likened to that of the copper wires in the electric circuit. Both offer little opposition to flow (so that a negligible fraction of the driving force (MMF or EMF) is wasted in conveying the flow to the load) and both can be shaped to guide the flow to its destination. There is one important difference, however. In the electric circuit, no current will flow until the circuit is completed, after which all the current is confined inside the wires. With an iron magnetic circuit, some flux can flow (in the surrounding air) even before the iron is installed. And although most of the flux will subsequently take the easy route through the iron, some will still leak into the air, as shown in Figure 1.4. We will not pursue leakage flux here, though it is sometimes important, as will be seen later.

Air-gap flux densities

If we neglect the reluctance of the iron parts of a magnetic circuit, it is relatively easy to estimate the flux density in the air-gap. Since the iron parts are then in effect 'perfect conductors' of flux, none of the source MMF (NI) is used in driving the flux through the iron parts, and all of it is available to push the flux across the air-gap. The situation depicted in Figure 1.4 therefore reduces to that shown in Figure 1.5, where an MMF of NI is applied directly across an air-gap of length g.

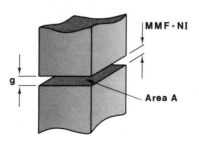

Figure 1.5 *Air-gap region, with MMF of NI applied across opposing polefaces*

The reluctance of any 'prism' of air, of cross-sectional area A and length g (Figure 1.5) is given by

$$\Lambda = \frac{g}{\mu_0 A} \tag{1.5}$$

where μ_0 is the magnetic space constant or permeability of free space which quantifies the magnetic properties of air. In SI units, $\mu_0 = 4\pi \times 10^{-7}$ H/m.

Equation 1.5 reveals the expected result that doubling the air-gap would double the reluctance (because the flux has twice as far to go), while doubling the area would halve the reluctance (because the flux has two equally appealing paths in parallel). To calculate the flux, Φ, we use the magnetic Ohm's law (equation 1.4), which gives

$$\Phi = \frac{\text{MMF}}{\Lambda} = \frac{NIA\mu_0}{g} \tag{1.6}$$

We are usually interested in the flux density in the gap, rather than the flux, so we use equation 1.1 to yield

$$B = \frac{\Phi}{A} = \frac{\mu_0 NI}{g} \tag{1.7}$$

Equation 1.7 is very simple, and from it we can calculate the air-gap flux density once we know the MMF of the coil (NI) and the length of the gap (g). For example, suppose the magnetising coil has 250 turns, the current is 2 A, and the gap is 1 mm. The flux density is then given by

$$B = \frac{4\pi \times 10^{-7} \times 250 \times 2}{1 \times 10^{-3}} = 0.63 \text{ tesla}$$

Note that we do not need to know the cross-sectional area of the magnetic circuit to obtain the gap flux density, which depends only on the length of the gap and the MMF.

If the cross-sectional area of the iron were constant at all points, the flux density would be 0.63 T everywhere. Sometimes, however, the cross-section of the iron reduces at points away from the air-gap, as shown for example in Figure 1.6.

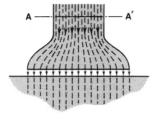

Figure 1.6   *Sketch showing increased flux density where the cross-sectional area of the magnetic circuit reduces*

Because the flux is compressed in the narrower sections, the flux density is higher, and in Figure 1.6 if the flux density at the air-gap and in the adjacent pole-faces is once again taken to be 0.63 T, then at the section AA' (where the area is only half that at the air-gap) the flux density will be 2 × 0.63 = 1.26 T.

Saturation

It would be reasonable to ask whether there is any limit to the flux density at which the iron can be operated. We can anticipate that there must be a limit, or else it would be possible to squash the flux into a vanishingly small cross-section, which we know from experience is not the case. In fact there is a limit, though not a very sharply defined one.

Earlier we noted that the iron has almost no reluctance, at least not in comparison with air. Unfortunately this happy state of affairs is only true as long as the flux density remains below about 1.6–1.8 T, depending on the particular steel in question. If we try to work the iron at higher flux densities, it begins to exhibit significant reluctance, and no longer behaves like an ideal conductor of flux. At these higher flux densities a significant proportion of the source MMF is used in driving the flux through the iron. This situation is obviously undesirable, since less MMF remains to drive the flux across the air-gap. So just as we would not recommend the use of high-resistance supply leads to the load in an electric circuit, we must avoid saturating the iron parts of the magnetic circuit.

The emergence of significant reluctance as the flux density is raised is illustrated qualitatively in Figure 1.7.

When the reluctance begins to be appreciable, the iron is said to be beginning to 'saturate'. The term is apt, because if we continue increasing the MMF, or reducing the area of the iron, we will eventually reach an almost constant flux density, typically around 2 T. To avoid the undesirable effects of saturation, the size of the iron parts of the magnetic circuit are usually chosen so that the flux density does not exceed about 1.5 T. At this level of flux density, the reluctance of the iron parts will be small in comparison with the air-gap.

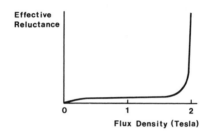

Figure 1.7   *Sketch showing how the effective reluctance of iron increases rapidly as the flux density approaches saturation*

Magnetic circuits in motors

The reader may be wondering why so much attention has been focused on the gapped C-core magnetic circuit, when it appears to bear little resemblance to the iron parts found in motors. We shall now see that it is actually a short step from the C-core to a motor magnetic circuit, and that no fundamentally new ideas are involved.

The evolution from C-core to motor geometry is shown in Figure 1.8, which should be largely self-explanatory, and relates to the field system of a d.c. motor.

We note that the first stage of evolution (Figure 1.8(b)) results in the original single gap of length g being split into two gaps of length $g/2$, reflecting the requirement for the rotor to be

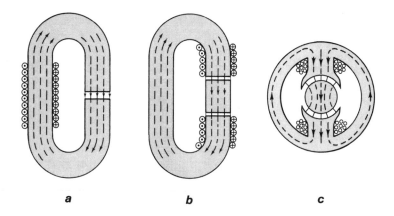

Figure 1.8   *Evolution of magnetic circuit of d.c. motor from simple C-core*

able to turn. At the same time the single magnetising coil is split into two to preserve symmetry. (Relocating the magnetising coil at a different position around the magnetic circuit is of course in order, just as a battery can be placed anywhere in an electric circuit.) Finally (Figure 1.8(c)), the single magnetic path is split into two parallel paths of half the original cross-section, each of which carries half of the flux, and the pole faces are curved to match the rotor. The coil now has several layers in order to fit the available space, but as discussed earlier this has no adverse effect on the MMF. The air-gap is still small, so the flux crosses radially to the rotor.

TORQUE PRODUCTION

Having designed the magnetic circuit to give a high flux density under the poles (Figure 1.8c), we must obtain maximum benefit from it. We therefore need to arrange a set of conductors on the rotor, as shown in Figure 1.9, and to ensure that conductors under a N-pole (at the top of Fig 1.9) carry positive current (into the paper), while those under the S-pole carry negative current. The force on all the positive conductors will be to the left, while the force on the negative ones will be to the right. A

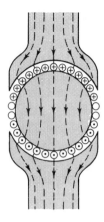

Figure 1.9   *Current-carrying conductors on rotor, positioned to maximise torque. The source of the magnetic flux lines (arrowed) is not shown*

nett couple, or torque will therefore be exerted on the rotor, which will be caused to rotate.

Magnitude of torque

The force on each conductor is given by equation 1.2, and it follows that the total tangential force F depends on the flux density produced by the field winding, the number of conductors on the rotor, the current in each, and the length of the rotor. The resultant torque (T) depends on the radius of the rotor (r), and is given by

$$T = Fr \qquad (1.8)$$

Slotting

If the conductors were mounted on the surface of the rotor iron, as in Figure 1.9, the air-gap would have to be at least equal to the wire diameter, and the conductors would have to be secured to the rotor in order to transmit their turning force to it. The earliest motors were made like this, with string or tape to bind the conductors to the rotor.

We can avoid the penalty of the large air-gap (which results in an unwelcome high reluctance in the magnetic circuit) by placing the conductors in slots in the rotor, as shown in Figure 1.10.

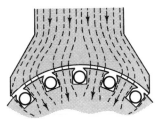

Figure 1.10   *Influence on flux paths when the rotor surface is slotted to accommodate conductors*

With the conductors in slots, the air-gap can be made small, but, as can be seen from Figure 1.10, almost all the flux now passes down the low-reluctance path through the teeth, leaving the conductors exposed to the very low leakage flux density in the slots. It might therefore be expected that little or no force would be developed, since on the face of it the conductors are screened from the flux. Remarkably, however, what happens is that the total force remains the same as it would have been if the conductors were actually in the flux, but almost all the force now acts on the rotor teeth, rather than on the conductors themselves.

This is very good news indeed. By putting the conductors in slots, we simultaneously reduce the reluctance of the magnetic circuit, and transfer the force to the rotor iron, which is robust and well able to transmit the resulting torque to the shaft. There are some snags, however. To maximise the torque, we will want as much current as possible in the rotor conductors. Naturally we will work the copper at the highest practicable current density (typically between 2 and 8 A/mm^2), but we will also want to maximise the cross-sectional area of the slots to accommodate as much copper as possible. This will push us in the direction of wide slots, and hence narrow teeth. But we recall that the flux has to pass radially down the teeth, so if we make the teeth too narrow, the iron in the teeth will saturate, and lead to a poor magnetic circuit. There is also the possibility of increasing the depth of the slots, but this cannot be taken too far or the centre region of the rotor iron will become so depleted that it too will saturate.

SPECIFIC LOADINGS AND SPECIFIC OUTPUT

Specific loadings

A design compromise is inevitable in the crucial air-gap region, and designers constantly have to exercise their skills to achieve the best balance between the conflicting demands on space made by the flux (radial) and the current (axial).

As in most engineering design, guidelines emerge as to what can be achieved in relation to particular sizes and types of machine, and motor designers usually work in terms of two parameters, the specific magnetic loading, and the specific electric loading. These parameters have a direct bearing on the output of the motor, as we will now see.

The specific magnetic loading $(\bar{B})$ is the average radial flux density over the cylindrical surface of the rotor, while the specific electric loading (usually denoted by the symbol $(\bar{A})$, the A standing for amperes) is the axial current per metre of circumference on the rotor. Many factors influence the values which can be employed in motor design, but in essence both parameters are limited by the properties of the materials (iron for the flux, and copper for the current), and by the cooling system employed to remove heat losses.

The specific magnetic loading does not vary greatly from one machine to another, because the saturation properties of most core steels are similar. On the other hand, quite wide variations occur in the specific electric loadings, depending on the type of cooling used.

Despite the low resistivity of the copper conductors, heat is generated by the flow of current, and the current must therefore be limited to a value such that the insulation is not damaged by an excessive temperature rise. The more effective the cooling

Plate 1.1 *Totally enclosed fan-ventilated (TEFV) cage induction motor. This particular example is rated at 200 W (0.27 h.p.) at 1450 rev/min, and is at the lower end of the power range for 3-phase versions. The case is of cast aluminium, with cooling air provided by the covered fan at the non-drive end. Note the provision for alternative mounting. (Photograph by courtesy of Brook Crompton)*

system, the higher the electric loading can be. For example, if the motor is totally enclosed and has no internal fan, the current density in the copper has to be much lower than in a similar motor which has a fan to provide a continuous flow of ventilating air. Similarly, windings which are fully impregnated with varnish can be worked much harder than those which are surrounded by air, because the solid body of encapsulating varnish provides a much better thermal path along which the heat can flow to the stator body. Overall size also plays a part in determining permissible electric loading, with larger motors generally having higher values than small ones.

In practice, the important point to be borne in mind is that unless an exotic cooling system is employed, most motors (induction, d.c. etc.) of a particular size have more or less the same specific loadings, regardless of type. As we shall now see, this in turn means that motors of similar size have similar torque capabilities. This fact is not widely appreciated by users, but is always worth bearing in mind.

Torque and motor volume

In the light of the earlier discussion, it follows that the tangential force per unit area of the rotor surface is equal to the product of the two specific loadings, i.e. $\bar{B}\bar{A}$. To obtain the total tangential force we must multiply by the area of the curved surface of the rotor, and to obtain the total torque we multiply the total force by the radius of the rotor.

Hence for a rotor of diameter D and length L, the total torque is given by

$$T \quad \alpha \quad \bar{B}\bar{A}D^2 L \tag{1.9}$$

This equation is extremely important. The term D^2L is proportional to the rotor volume, so we see that for given values of the specific magnetic and electric loadings, the torque from any motor is proportional to the rotor volume. We are at liberty to choose a long thin rotor or a short fat one, but once the rotor volume and specific loadings are specified, we have effectively determined the torque.

It is worth stressing that we have not focused on any particular type of motor, but have approached the question of torque production from a completely general viewpoint. In essence our conclusions reflect the fact that all motors are made from iron and copper, and differ only in the way these materials are disposed. We should also acknowledge that in practice it is the overall volume of the motor which is important, rather than the volume of the rotor. But again we find that, regardless of the type of motor, there is a fairly close relationship between the overall volume and the rotor volume, for motors of similar torque. We can therefore make the bold but generally accurate statement that the overall volume of a motor is determined by the torque it has to produce. There are of course exceptions to this rule, but as a general guideline for motor selection, it is extremely useful.

Having seen that torque depends on volume, we must now turn our attention to the question of power output.

Specific output power – importance of speed

The work done by a motor delivering a torque T is equal to the torque times the angle turned through, and the power, which is the rate of working, is therefore equal to the torque times the angular speed (ω), i.e.

$$P = T\omega \qquad (1.10)$$

We can now express the power output in terms of the rotor dimensions and the specific loadings, using equation 1.9 which yields

$$P \quad \alpha \quad \bar{B}\bar{A}D^2L\omega \qquad (1.11)$$

Equations 1.10 and 1.11 emphasise the importance of speed in determining power output. For a given power, we can choose between a large (and therefore expensive) low-speed motor or a small (and cheaper) high-speed one. The latter choice is preferred for most applications, even if some form of speed reduction (using belts or gears, for example) is needed, because the smaller motor is cheaper. Familiar examples include portable electric tools, where rotor speeds of 12 000 rev/min or more allow powers of hundreds of watts to be obtained, and electric traction, where the motor speed is considerably higher than the wheel speed. In these examples, where volume and weight are at a premium, a direct drive would be out of the question.

The significance of speed is underlined when we rearrange equation 1.11 to obtain an expression for the specific power output (power per unit rotor volume), Q, given by

$$Q \quad \alpha \quad \bar{B}\bar{A}\omega \qquad (1.12)$$

To obtain the highest possible specific output for given values of the specific magnetic and electric loadings, we must clearly operate the motor at the highest practicable speed. The one obvious disadvantage of a small high-speed motor and gearbox is that the acoustic noise (both from the motor itself and from the power transmission) is higher than it would be from a larger direct drive motor. When noise must be minimised (for example in ceiling fans), a direct drive motor is therefore preferred, despite its larger size.

MOTIONAL EMF

We have already seen that force (and hence torque) is produced on current-carrying conductors exposed to a magnetic field. The force is given by equation 1.2, which shows that as long as the flux density and current remain constant, the force will be constant. In particular we see that the force does not depend on whether the conductor is stationary or moving. On the other hand relative movement is an essential requirement in the production of mechanical output power (as distinct from torque), and we have seen that output power is given by the equation $P = T\omega$. We will now see that the presence of relative motion between the conductors and the field always brings 'motional e.m.f.' into play; and we shall see that this motional e.m.f. is an essential feature of the energy conversion process.

Power relationships – stationary conditions

We look first at the electrical input power, considering the elementary set-up depicted in Figure 1.11, and beginning with the conductor stationary (i.e. $v = 0$).

For the purpose of this discussion we can suppose that the magnetic field (B) is provided by a permanent magnet. Once the field has been established (when the magnet was first magnetised and placed in position), no further energy will be needed to sustain the field, which is just as well since it is obvious that an inert magnet is incapable of continuously supplying energy. It follows that when we obtain mechanical output from this primitive 'motor', none of the energy involved comes from the magnet.

When the conductor is held stationary the force produced on

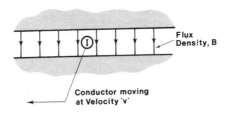

Figure 1.11   *Elementary 'motor' showing conductor carrying current* I *(perpendicular to the plane of the paper), and moving with velocity* v *perpendicular to the magnetic flux density,* B

it (BIl) does no work, so there is no mechanical output power, and the only electrical input power required is that needed to drive the current through the conductor. If the resistance of the conductor is R, and the current through it is I, the voltage required will be given by $V_1 = IR$, and the electrical input power will be $V_1 I$ or $I^2 R$. Under these conditions, all the electrical input power will appear as heat inside the conductor, and the power balance can be expressed by the equation

electrical input power $(V_1 I)$ = rate of loss of heat in
                        conductor $(I^2 R)$ (1.13)

Power relationships – conductor moving at constant speed

Now let us imagine the situation where the conductor is allowed to move at a constant velocity (v) in the direction of the force, and that the current in the conductor is maintained at the value (I) which it had when it was stationary. We need not worry at this stage about the practicalities of being able to supply current to a moving conductor; the important point is that the situation is typical of what happens in a real machine.

Mechanical work is now being done against the opposing force of the load, and the mechanical output power is equal to the rate of work, i.e. the force (BIl) times the velocity (v). The power lost as heat in the conductor is the same as it was when stationary, since it has the same resistance, and the same current. The electrical input power must continue to supply this heat loss, but in addition it must now furnish the mechanical output power. The power balance equation now becomes

$$\text{electrical input power} = \text{rate of loss of heat in conductor}$$
$$+ \text{ mechanical output power} \qquad (1.14)$$
$$= I^2R + BIlv$$

Clearly, we would expect the electrical input power to be greater than it was when the conductor was stationary, and we can see from the second term in equation 1.14 that it must increase directly with the speed. The question which then springs to mind is how does the source supplying current to the conductor know that it has to supply more power in proportion to the speed of the conductor? There must be some electrical effect which accompanies the motion, and causes a change in the way the conductor appears to the supply, otherwise the situation on the electrical side would be unchanged.

The key to unlock the question lies in our assumption that the current is constant all the time. Electrical input power is equal to current times voltage, so since the current is assumed to have been kept constant it follows that the voltage of the source must increase from the value it had when the conductor was stationary, by an amount sufficient to furnish the extra mechanical power. If we denote the required source voltage when the conductor is moving by V_2, the power balance equation can be written

$$V_2I = I^2R + BIlv \qquad (1.15)$$

Hence by combining equations 1.13 and 1.15 we obtain

$$(V_2 - V_1)I = BlvI \qquad (1.16)$$

and thus

$$V_2 - V_1 = E = Blv \qquad (1.17)$$

Equation 1.16 shows the expected result that the increase in source power is equal to the mechanical output power, while equation 1.17 quantifies the extra voltage to be provided by the source to keep the current constant when the conductor is moving. This increase in source voltage is a reflection of the fact that whenever a conductor moves through a magnetic field, an e.m.f. (E) is induced in it.

We see from equation 1.17 that the e.m.f. is directly proportional to the flux density, to the velocity of the conductor relative to the flux, and to the length of the conductor. The source voltage has to overcome this additional voltage in order to keep the same current flowing: if the source voltage were not increased, the current would fall as soon as the conductor began to move because of the opposing effect of the induced e.m.f.

We have deduced that there must be an e.m.f. caused by the motion, and have derived an expression for it by using the principle of the conservation of energy, but the result we have obtained, i.e.

$$E = Blv \qquad (1.18)$$

is often introduced as the 'flux-cutting' form of Faraday's law, which states that when a conductor moves through a magnetic field an e.m.f. given by equation 1.18 is induced in it. Because motion is an essential part of this mechanism, the e.m.f. induced is referred to as a 'motional e.m.f.'. The 'flux-cutting' terminology arises from attributing the origin of the e.m.f. to the cutting or slicing of the lines of flux by the passage of the conductor. This is a useful mental picture, though it must not be pushed too far: the flux lines are after all merely inventions which we find helpful in coming to grips with magnetic matters.

Before turning to the equivalent circuit of the primitive motor two general points are worth noting. Firstly, whenever energy is being converted from electrical to mechanical form, as here, the induced e.m.f. always acts in opposition to the applied (source) voltage. This is reflected in the use of the term 'back e.m.f.' to describe motional e.m.f. in motors. Secondly, although we have discussed a particular situation in which the conductor carries current, it is certainly not necessary for any current to be flowing in order to produce an e.m.f.: all that is needed is relative motion between the conductor and the magnetic field.

Equivalent circuit

We can represent the electrical relationships in the moving conductor experiment in an equivalent circuit as shown in Figure 1.12.

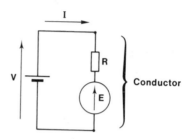

Figure 1.12   *Equivalent circuit for the elementary motor shown in Figure 1.11*

The motional e.m.f. in Figure 1.12 is shown as opposing the applied voltage, which applies in the 'motoring' condition we have been discussing. Applying Kirchhoff's law we obtain the voltage equation as

$$V = E + IR \quad \text{or} \quad I = \frac{V - E}{R} \tag{1.19}$$

Multiplying equation 1.19 by the current gives the power equation as

electrical input power (VI) = *mechanical output power (EI)*

+ copper loss (I^2R) $\qquad$ (1.20)

(Note that the term 'copper loss' used in equation 1.20 refers to the heat generated by the current in the windings: all such losses in electric motors are referred to in this way, even when the conductors are made of aluminium or bronze!)

It is worth seeing what can be learned from these equations because, as noted earlier, this simple elementary 'motor' encapsulates all the essential features of real motors. Lessons which emerge at this stage will be invaluable later, when we look at the way actual motors behave.

If the e.m.f. E is less than the applied voltage V, the current will be positive, and electrical power will flow from the source, resulting in motoring action. On the other hand if E is larger than V, the current will flow back to the source, and the conductor will be acting as a generator. This inherent ability to switch from motoring to generating without any interference by the user is an extremely desirable property of electromagnetic energy converters. Our primitive set-up is simply a machine which is equally at home acting as motor or generator.

A further important point to note is that the mechanical power (the first term on the right hand side of equation 1.20) is simply the motional e.m.f. multiplied by the current. This result is again universally applicable, and easily remembered. We may sometimes have to be a bit careful if the e.m.f. and the current are not simple d.c. quantities, but the basic idea will always hold good.

Motoring condition

Motoring implies that the conductor is moving in the same direction as the electromagnetic force (BIl), and at a speed such that the back e.m.f. (BLv) is less than the applied voltage V. In the discussion so far, we have assumed that the applied voltage is adjusted so that the current is kept constant. This was a helpful approach to take in order to derive the steady-state power relationships, but is seldom typical of normal operation. We therefore turn to how the moving conductor will behave under conditions where the applied voltage V is constant, since this corresponds more closely with operation of a real motor.

Behaviour with no mechanical load

If we begin with the conductor stationary when the voltage V is first applied, the current will immediately rise to a value of V/R, since there is no motional e.m.f. and the only thing which limits the current is the resistance. (Strictly we should allow for the effect of inductance in delaying the rise of current, but we choose to ignore it here in the interests of simplicity.) The current will be large, and a high force will therefore be developed on the conductor. If there are no other forces acting, the conductor will have a high acceleration in the direction of the force. As it picks up speed, the motional e.m.f. (equation 1.18) will grow in proportion to the speed. Since the motional e.m.f. opposes the applied voltage, the current will fall (equation 1.19), so the force and hence the acceleration will reduce, though the speed will continue to rise. The speed will continue to increase as long as there is an accelerating force, i.e. as long as there is a current in the conductor. We can see from equation 1.19 that the current will finally fall to zero when the speed reaches a level at which the motional e.m.f. is equal to the applied voltage. The speed and current therefore vary as shown in Figure 1.13, both curves having the exponential shape which characterises the response of systems governed by a first-order differential equation.

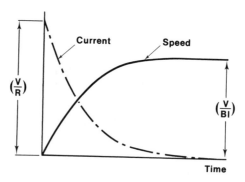

Figure 1.13   *Dynamic behaviour of the elementary motor with no mechanical load*

We note that in this idealised situation (in which there is no load applied, and no friction forces), the conductor will continue to travel at a constant speed, because with no nett force acting on it there is no acceleration. Of course, no mechanical power is being produced, since we have assumed that there is no opposing force on the conductor, and there is no input power because the current is zero. This hypothetical situation nevertheless corresponds closely to the so-called no-load condition in a motor, the only difference being that a motor will have some friction, which has been ignored here in order to simplify the discussion.

An elegant self-regulating mechanism is evidently at work here. When the conductor is stationary, it has a high force acting on it, but this force tapers off as the speed rises to its target value, which corresponds to the back e.m.f. being equal to the applied voltage. Looking back at the expression for motional e.m.f. (equation 1.18), we can obtain an expression for the no-load speed v_0 by equating the applied voltage and the back e.m.f., which gives

$$V = E = Blv_0 \qquad \therefore v_0 = \frac{V}{Bl} \qquad (1.21)$$

Equation 1.21 shows that the steady-state no-load speed is directly proportional to the applied voltage, which indicates that speed control can be achieved by means of the applied voltage.

Rather more surprisingly, however, the speed is seen to be inversely proportional to the magnetic flux density, which means that the weaker the field, the higher the steady-state speed. This result can cause raised eyebrows, and with good reason. Surely, it is argued, since the force is produced by the action of the field, the conductor will not go as fast if the field is weaker. This view is wrong, but understandable. The flaw in the argument is to equate force with speed. When the voltage is first applied, the force on the conductor certainly will be less if the field is weaker, and the initial acceleration will be lower. But in both cases the acceleration will continue until the current has fallen to zero, and this will only happen when the induced e.m.f. has risen to equal the applied voltage. With a weaker field, the speed needed to generate this e.m.f. will be higher than with a strong field: there is 'less flux', so what there is has to be cut at a higher speed to generate a given e.m.f. The matter is summarised in Figure 1.14, which shows how the speed will rise for a given applied voltage, for 'full' and 'half' fields respectively. Note that the initial acceleration (i.e. the slope of the speed–time curve) in the half-flux case is half that of the full flux case, but the final steady speed is twice as high.

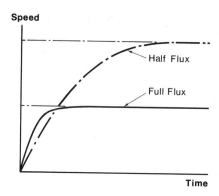

Figure 1.14   *Effect of flux on the acceleration and steady running speed of the elementary motor*

Behaviour with a mechanical load

Suppose we now apply a steady force (the load) opposing the motion of the conductor. The conductor will begin to decelerate, but as soon as the speed falls, the back e.m.f. will become less than V, and current will begin to flow. The more the speed drops, the bigger the current, and hence the more the electromagnetic force developed by the conductor. When the force developed by the conductor becomes equal to the force we have applied, the deceleration will cease, and a new equilibrium condition will be reached. The speed will be lower than at no-load, and the conductor will now be producing continuous mechanical output power, i.e. acting as a motor.

Since the electromagnetic force on the conductor is directly proportional to the current, it follows that the steady-state current is directly proportional to the load which is applied. We note from equation 1.19 that the current depends directly on the difference between V and E, and inversely on the resistance. Hence for a given resistance, the larger the load (and hence the steady-state current), the greater the required difference between V and E, and hence the lower the steady running speed, as shown in Figure 1.15.

We can also see from equation 1.19 that the higher the resistance of the conductor, the more it slows down when a

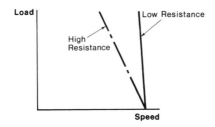

Figure 1.15   *Influence of resistance on the drop in speed with load*

given load is applied. Conversely, the lower the resistance, the more the conductor is able to hold its no-load speed in the face of applied load. This is also illustrated in Fig 1.15. We can deduce that the only way we could obtain an absolutely constant speed with this type of motor is for the resistance of the conductor to be zero, which is of course not possible.

We complete our exploration of the performance when loaded by asking how the flux density influences behaviour. Recalling that the electromagnetic force is proportional to the flux density as well as the current, we can deduce that to develop a given force, the current required will be higher with a weak flux than with a strong one. Hence in view of the fact that there will always be an upper limit to the current which the conductor can safely carry, the maximum force which can be developed will vary in direct proportion to the flux density, with a weak flux leading to a low maximum force and vice-versa. This underlines the importance of operating with maximum flux density whenever possible.

We can also see another disadvantage of having a low flux density by noting that to achieve a given force, the drop in speed will be disproportionately high when we go to a lower flux density. We can see this by imagining that we want a particular force, and considering how we achieve it firstly with full flux, and secondly with half flux. With full flux, there will be a certain drop in speed which causes the motional e.m.f. to fall enough to admit the required current. But with half the flux, for example, twice as much current will be needed to develop the same force. Hence the motional e.m.f. must fall by twice as much as it did with full flux. However, since the flux density is now only half, the drop in speed will have to be four times as great as it was with full flux. The half-flux 'motor' therefore has a load characteristic with a load/speed gradient four times more droopy than the full-flux one. This is shown in Figure 1.16, the applied voltages having been adjusted so that in both cases the no-load speed is the same. The half-flux motor is clearly inferior in terms of its ability to hold the set speed when the load is applied.

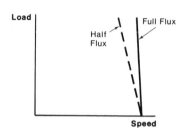

Figure 1.16   *Influence of flux on the drop in speed with load*

We may be tempted to think that the higher speed which we can obtain by reducing the flux somehow makes for better performance, but we can now see that this is not so. By halving the flux, for example, the no-load speed for a given voltage is doubled, but when the load is raised until rated current is flowing in the conductor, the force developed is only half, so the mechanical power is the same. We are in effect trading speed against force, and there is no suggestion of getting something for nothing.

Relative magnitudes of V and E, and efficiency

Invariably we want machines which have high efficiency. From equation 1.20, we see that to achieve high efficiency, the copper loss (I^2R) must be small compared with the mechanical power (EI), which means that the resistive volt-drop in the conductor (IR) must be small compared with either the induced e.m.f. (E) or the applied voltage (V). In other words we want most of the applied voltage to be accounted for by the 'useful' motional e.m.f., rather than the wasteful volt drop in the wire. Since the motional e.m.f. is proportional to speed, and the resistive volt drop depends on the conductor resistance, we see that a good energy converter requires the conductor resistance to be as low as possible, and the speed to be as high as possible.

To provide a feel for the sorts of numbers likely to be encountered, we can consider a conductor with resistance of $0.5\,\Omega$, carrying a current of 4 A, and moving at a speed such that the motional e.m.f. is 8 V. From equation 1.19, the supply voltage is given by

$$V = E + IR = 8 + 4 \times 0.5 = 10 \text{ volts}$$

Hence the electrical input power (VI) is 40 watts, the mechanical output power (EI) is 32 watts, and the copper loss (I^2R) is 8 watts, giving an efficiency of 80%.

If the supply voltage is doubled (i.e. $V = 20$ volts), however, and the resisting force is assumed to remain the same (so that the steady-state current is still 4 A), the motional e.m.f. is given by equation 1.19 as

$$E = 20 - 4 \times 0.5 = 18 \text{ volts}$$

which shows that the speed will have rather more than doubled, as expected. The electrical input power is now 80 watts, the mechanical output power is 72 watts, and the copper loss is still 8 watts. The efficiency has now risen to 90%, underlining the fact that the energy conversion process gets better at higher speeds.

The ideal situation is clearly one where the term IR in equation 1.19 is negligible, so that the back e.m.f. is equal to the applied voltage. We would then have an ideal machine with an efficiency of 100%, in which the steady-state speed would be directly proportional to the applied voltage and independent of the load.

In practice the extent to which we can approach the ideal situation discussed above depends on the size of the machine. Tiny motors, such as those used in wrist-watches, are awful, in that most of the applied voltage is used up in overcoming the resistance of the conductors, and the motional e.m.f. is very small: these motors are much better at producing heat than they are at producing mechanical output power! Small machines, such as those used in hand tools, are a good deal better with the motional e.m.f. accounting for perhaps 70–80% of the applied voltage. Industrial machines are very much better: the largest ones (of many hundreds of kW) use only one or two per cent of the applied voltage in overcoming resistance, and therefore have very high efficiencies.

GENERAL PROPERTIES OF ELECTRIC MOTORS

All electric motors are governed by the laws of electromagnetism, and are subject to essentially the same constraints

Plate 1.2   *Steel frame cage induction motor, 150 kW (201 h.p.), 1485 rev/min. The active parts are totally enclosed, and cooling is provided by means of an internal fan which circulates cooling air round the interior of the motor through the hollow ribs, and an external fan which blows air over the case. (Photograph by courtesy of Brook Crompton)*

imposed by the materials (copper and iron) from which they are made. We should therefore not be surprised to find that at the fundamental level all motors – regardless of type – have a great deal in common.

These common properties, most of which have been touched on in this chapter, are not usually given prominence. Books tend to concentrate on the differences between types of motor, and manufacturers are usually interested in promoting the virtues of their particular motor at the expense of the competition. This divisive emphasis causes the underlying unity to be obscured, leaving users with little opportunity to absorb the sort of knowledge which will equip them to make informed judgements.

The most useful ideas worth bearing in mind are therefore given below, with brief notes accompanying each. Experience indicates that users who have these basic ideas firmly in mind will find themselves able to understand why one motor seems better than another, and will feel much more confident when faced with the difficult task of weighing the pros and cons of competing types.

Operating temperature and cooling

The cooling arrangement is the single most important factor in determining the output from any given motor.

Any motor will give out more power if its electric circuit is worked harder (i.e. if the current is allowed to increase). The limiting factor is normally the allowable temperature rise of the windings, which depends on the class of insulation.

For class F insulation (the most widely used) the permissible temperature rise is 100 K, whereas for class H it is 125 K. Thus if the cooling remains the same, more output can be obtained simply by using the higher-grade insulation. Alternatively, with a given insulation the output can be increased if the cooling system is improved. A through-ventilated motor, for example, might give perhaps twice the output power of an otherwise identical but totally enclosed machine.

Torque per unit volume

*For motors with similar cooling systems, the rated torque is
approximately proportional to the rotor volume, which in turn
is roughly proportional to the overall motor volume.*

This stems from the fact that for a given cooling arrangement, the specific and magnetic loadings of machines of
different types will be more or less the same. The torque per unit
length therefore depends first and foremost on the square of the
diameter, so motors of roughly the same diameter and length
can be expected to produce roughly the same torque.

Power per unit volume – importance of speed

Output power per unit volume is directly proportional to speed.

Low-speed motors are unattractive because they are large, and
therefore expensive. It is usually much better to use a high-
speed motor with a mechanical speed reduction. For example,
a direct drive motor for a portable electric screwdriver would
be an absurd proposition.

Size effects – specific torque and efficiency

*Large motors have a higher specific torque (torque per unit
volume) and are more efficient than small ones.*

In large motors the specific electric loading is normally much
higher than in small ones, and the specific magnetic loading is
somewhat higher. These two factors combine to give the higher
specific torque.

Very small motors are inherently very inefficient (e.g. 1% in a
wrist-watch), whereas motors of over say 100 kW (134 h.p.)
have efficiencies above 95%. The reasons for this scale effect are
complex, but stem from the fact that the resistance volt-drop
term can be made relatively small in large electromagnetic
devices, whereas in small ones the resistance becomes the
dominant term.

Efficiency and speed

The efficiency of a motor improves with speed.

For a given torque, power output rises in proportion to speed, while electrical losses are – broadly speaking – constant. Efficiency therefore rises with speed.

Rated voltage

A motor can be provided to suit any voltage.

Within limits it is always possible to rewind a motor for a different voltage without affecting its performance. A 200 V, 10 A motor could be rewound for 100 V, 20 A simply by using half as many turns per coil of wire having twice the cross-sectional area. The total amounts of active material, and hence the performance, would be the same.

Short-term overload

Most motors can be overloaded for short periods without damage.

The continuous electric loading (i.e. the current) cannot be exceeded without damaging the insulation, but if the motor has been running with reduced current for some time, it is permissible for the current (and hence the torque) to be much greater than normal for a short period of time. The principal factors which influence the magnitude and duration of the permissible overload are the thermal time-constant (which governs the rate of rise of temperature) and the previous pattern of operation. Thermal time constants range from a few seconds for small motors to many minutes or even hours for large ones. Operating patterns are obviously very variable, so rather than rely on a particular pattern being followed, it is usual for motors to be provided with over-temperature protective devices (e.g. thermistors) which trigger an alarm and/or trip the supply if the safe temperature is exceeded.

QUESTIONS ARISING

- *We have heard a lot about the torque on the rotor, but presumably there must also be a torque on the stator. Can the mechanism of torque on the stator be pictured using the same ideas as for the rotor torque?*

Yes, there is always an equal and opposite torque on the stator, which is why it is usually important to bolt a motor down securely. In some machines (e.g. the induction motor) it is easy to see that torque is produced on the stator by the interaction of the air-gap flux density and the stator currents, in exactly the same way that the flux density interacts with the rotor currents to produce torque on the rotor. In other motors (e.g. the d.c. motor), there is no simple physical argument which can be advanced to derive the torque on the stator, but nevertheless it is equal and opposite to the torque on the rotor.

- *In the discussion of how an elementary motor works, there is no mention of the magnetic field produced by the current-carrying conductor(s) on the rotor. Won't this field modify the field in the air-gap, and if so won't this invalidate the expression for the force on the conductor?*

This is a very perceptive question. The answer is that the field produced by the current-carrying conductor on the rotor will certainly modify the original field (i.e. the field that was present when there was no current in the rotor conductor). But in the majority of motors, the force on the conductor can be calculated correctly from the product of the current and the 'original' field. This not only makes force calculation relatively easy, but also has a logical feel to it. For example in Figure 1.11, we would not expect any force on the current-carrying conductor if there were no externally applied field, even though the current in the conductor will produce its own field (upwards on one side of the conductor and downwards on the other). So it seems right that since we only obtain a force when there is an external field, all of the force must be due to that field alone.

- *When I look at a diagram showing magnetic flux across an air-gap between two iron surfaces it reminds me of a lifting electromagnet, and I would expect there to be quite a strong magnetic pull between the two surfaces. In the case shown in Fig 1.10, for example, I would expect the rotor to be attracted to the stator pole. Is there such a force in motors, and if so what prevents the rotor from being pulled sideways?*

There is indeed a radial force due to magnetic attraction, exactly as in a lifting magnet or relay. The force per unit area of pole-face is proportional to the square of the radial flux density, and with typical air-gap flux densities of up to 1 T in motors, the force per unit area of rotor surface works out to be about 40 N/cm^2. This indicates that the total radial force can be very large: for example the force of attraction on a small pole face of only 5 cm × 10 cm is 2000 N, or about 200 kg. This force contributes nothing to the torque of the motor, and is merely an unwelcome by-product of the 'BIl' mechanism we employ to produce tangential force on the rotor conductors.

It turns out that the radial magnetic force is actually a good deal bigger than the tangential electromagnetic force on the rotor conductors, and as the question implies, it tends to pull the rotor onto the pole. However, the majority of motors are constructed with an even number of poles equally spaced around the rotor, and the flux density in each pole is the same, so that – in theory at least – the resultant force on the complete rotor is zero. In practice, even a small eccentricity will cause the field to be stronger under the poles where the air-gap is smaller, and this will give rise to an unbalanced pull, resulting in noisy running and rapid bearing wear.

- *If the field windings in a motor do not contribute to the mechanical output power, why do they consume power continuously?*

The field windings act as the catalyst in the energy-conversion process, and as such they do not contribute to the mechanical output of the motor. But all windings have some resistance (R), so in order to drive a current (I) there has to be a voltage (V) such that $V = IR$. There will therefore be a steady input power of $VI(= I^2R)$, all of which appears as heat in the conductors forming the winding. The only ways to avoid this steady power loss are (a) to replace the field winding with a permanent magnet, or (b) to wind the coil from a superconducting material which has zero resistance. Option (b) will only be feasible if and when room temperature superconductors become available.

- *For a given power which will be larger, a motor or a generator?* Assuming that the speed of rotation is the same for both, the answer is neither. A given machine will work as well both as motor or generator. There may be detailed differences in the losses which have a marginal effect on the rated power, but to a first order motor and generator performance are the same.

- *I plan to replace the 440 V fixed-speed induction motor on a machine tool with a 110 V one, so that it can be used safely on a building site. Will the lower-voltage version fit in the existing space, and will the same control gear be usable?* Assuming that the new motor has the same rated power and speed, the 110 V version will be no different externally from the 440 V one. (All the stator coils will have a quarter of the number of turns of wire with four times the cross-sectional area, and the rotor will be the same – see Chapter 6.) Since it is a fixed-speed installation, the 'control gear' will probably amount to little more than an on/off contactor; the new motor will draw four times as much current, so it may be necessary to fit the contactor with beefier contacts.

- *I accept that for a given power, a high-speed motor will be smaller than a low-speed one. But most of the high-speed motors I have seen are noisy. How can I obtain a large, low-speed, quiet motor?* Unless you want to buy in very large quantities it is very unlikely that any manufacturer will be able to supply you. Recent emphasis has been on maximising the power/volume ratio, so speeds have tended to rise over the years. Perhaps as environmental factors assume greater importance, some enterprising manufacturer will see merit in producing motors of the strong silent type.

2

POWER ELECTRONIC CONVERTERS FOR MOTOR DRIVES

INTRODUCTION

In this chapter we will look at examples of the power converter circuits which are used with motor drives, providing either d.c. or a.c. outputs, and working from either a d.c. (battery) supply, or from the conventional a.c. mains. The treatment is not intended to be exhaustive, but should serve to highlight the most important aspects which are common to all types of drive converter.

Although there are many different types of converter, all except very low power ones are based on some form of electronic switching. The need to adopt a switching strategy is emphasised in the first example, where the consequences are explored in some depth. We shall see that switching is essential in order to achieve high-efficiency power conversion, but that the resulting waveforms are inevitably less than ideal from the point of view of the motor.

The examples have been chosen to illustrate typical practice, so for each the most commonly used switching devices (e.g. thyristor, transistor) are shown. In many cases, several different switching devices may be suitable (see later), so we should not identify a particular circuit as being the exclusive province of a particular device.

Before discussing particular circuits it will be useful to take an overall look at a typical drive system, so that the role of the converter can be seen in its proper context.

General arrangement of drive

A complete drive system is shown in block diagram form in Figure 2.1.

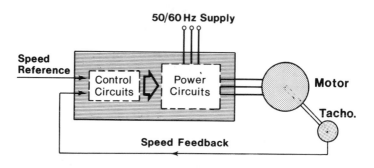

Figure 2.1 *General arrangement of speed-controlled drive*

The job of the converter is to draw electrical energy from the mains (at constant voltage and frequency) and supply electrical energy to the motor at whatever voltage and frequency is necessary to achieve the desired mechanical output.

Except in the very simplest converter (such as a simple diode rectifier), there are usually two distinct parts to the converter. The first is the power stage, through which the energy flows to the motor, and the second is the control section, which regulates the power flow. Control signals, in the form of low-power analogue or digital voltages, tell the converter what it is supposed to be doing, while other low-power feedback signals are used to measure what is actually happening. By comparing the demand and feedback signals, and adjusting the output accordingly, the target output is maintained. The simple arrangement shown in Figure 2.1 has only one input representing the desired speed, and one feedback signal indicating actual speed, but most drives will have extra feedback signals as we will see later.

A characteristic of power electronic converters which is shared with most electrical systems is that they have very little capacity for storing energy. This means that any sudden change in the power supplied by the converter to the motor must be reflected in a sudden increase in the power drawn from the supply. In most cases this is not a serious problem, but it does have two drawbacks. Firstly, sudden changes in the current drawn from the supply will cause spikes in the supply voltage, because of the effect of the supply impedance. These spikes will appear as unwelcome distortion to other users on the same supply. And secondly, there may be an enforced delay before the supply can react. With a single-phase mains supply, for example, there can be no sudden increase in the power supply from the mains at the instant where the mains voltage is passing through zero.

It would be better if a significant amount of energy could be stored within the converter itself: short-term energy demands could then be met instantly, thereby reducing rapid fluctuations in the power drawn from the mains. But unfortunately this is just not economic: most converters do have a small store of energy in their smoothing inductors and capacitors, but the amount is not sufficient to buffer the supply sufficiently to shield it from anything more than very short-term fluctuations.

VOLTAGE CONTROL – D.C. OUTPUT FROM D.C. SUPPLY

For the sake of simplicity we will begin with the problem of controlling the voltage across a resistive load, fed from a battery. Three different methods are shown in Figure 2.2. The battery voltage is assumed to be constant at 12 V, and we seek to vary the load voltage from 0 to 12 V. Although this is not quite the same as if the load were d.c. motor, the conclusions which we draw are effectively the same.

Method (a) uses a variable resistor (R) to absorb whatever fraction of the battery voltage is not required at the load. It provides smooth control, but the snag is that power is wasted in the control resistor. For example, if the load voltage is to be reduced to 6 V, the resistor R must be set to 2Ω, so that half of the battery voltage is dropped across R. The current will be 3 A, the load power will be 18 W, and the power dissipated in R will

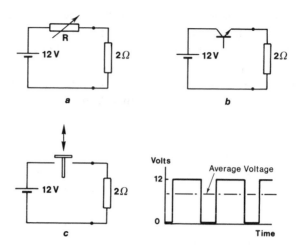

Figure 2.2   *Methods of obtaining variable-voltage d.c. output from constant-voltage source*

also be 18 W, giving an overall efficiency of only 50%. If R is increased further, the efficiency falls still lower, approaching zero as the load voltage tends to zero. This method of control is therefore unacceptable for motor control, except perhaps in low power applications such as car wiper and heater motors.

Method (b) is much the same as (a) except that a transistor is used instead of a manually-operated variable resistor. A transistor is a variable resistor, of course, but one in which the collector-emitter resistance can be controlled over a wide range by means of the base-emitter current. (The base-emitter current is usually very small, so it can be varied by means of a low-power electronic circuit whose losses are negligible in comparison with the power in the main (collector-emitter) circuit.)

The drawback of method (b) is the same as in (a) above, i.e. the efficiency is very low. But here the 'wasted' power (up to a maximum of 18 W in the example) is burned off inside the transistor which therefore has to be large, well-cooled, and hence expensive. Transistors are hardly ever operated in this 'linear' way when used in power electronics, but are widely used as switches, as discussed below.

Switching control

The basic ideas underlying a switching power regulator are shown by the arrangement in Figure 2.2(c), which uses a mechanical switch. By operating the switch repetitively and varying the ratio of 'on' time to 'off' time, the average load voltage can be varied smoothly between 0 V (switch off all the time) through 6 V (switch on and off for half of each cycle) to 12 V (switch on all the time).

The circuit shown in Figure 2.2(c) is often referred to as a 'chopper', because the battery supply is 'chopped' on and off. When a constant repetition frequency is used, and the width of the 'on' pulse is varied to control the mean output voltage, the arrangement is known as 'pulse width modulation' (PWM). An alternative approach is to keep the width of the 'on' pulses constant, but vary their repetition rate, and this is known as pulse frequency modulation.

The main advantage of the chopper circuit is that no power is wasted, and the efficiency is thus 100%. When the switch is 'on', current flows through it, but the voltage across it is zero because its resistance is negligible. The power dissipated in the switch is therefore zero. Likewise, when 'off', the current is zero, so although the voltage across the switch is 12 V, the power dissipated in it is again zero.

The disadvantage is that the load voltage waveform is no longer steady: it consists of a mean 'd.c.' level, with a superimposed 'a.c.' component. Bearing in mind that we really want the load to be a d.c. motor, rather than a resistor, we are bound to ask whether the pulsating voltage will be acceptable. The answer is yes, provided the frequency is high enough. If we pulse the switch at too low a rate, we will find that the speed of the motor will fluctuate in sympathy. But as the pulse rate is increased (keeping the same mark–space ratio), the speed ripple will reduce until it is imperceptible.

Obviously a mechanical switch would be inconvenient, and could not be expected to last long when pulsed at high frequency. So an electronic power switch is used instead. For the sake of simplicity, we will assume that the switching device is a conventional bipolar transistor (BJT). If we use a different switching device such as a metal oxide semiconductor field effect transistor (MOSFET) or an insulated gate bipolar transistor (IGBT) the detailed arrangements for turning the device on and off will be different (see later), but the main conclusions we draw will be much the same.

Transistor chopper

As noted earlier, a transistor is effectively a controllable resistor, i.e. the resistance between collector and emitter depends on the current in the base-emitter junction. In order to mimic the operation of a mechanical switch, the transistor would have to be able to provide infinite resistance (corresponding to an open switch) or zero resistance (corresponding to a closed switch). Neither of these ideal states can be reached with a real transistor, but both can be closely approximated.

The transistor will be 'off' when the base-emitter current is zero. Viewed from the main (collector-emitter) circuit, its resistance will be very high, as shown by the region Oa in Figure 2.3.

Under this 'cut-off' condition, only a tiny current will flow from collector to emitter, and the power dissipated will be negligible, giving an excellent approximation to an open switch.

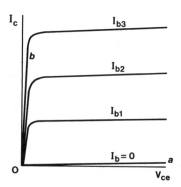

Figure 2.3   *Transistor characteristics showing high resistance (cut-off) region Oa, and low resistance (saturation) region Ob*

To turn the transistor fully 'on', a base-emitter current must be provided. The base current required will depend on the prospective collector-emitter current, i.e. the current in the load. The aim is to keep the transistor 'saturated' so that it has a very low resistance, corresponding to the region Ob in Figure 2.3. Typically in a bipolar transistor (BJT) the base current will need to be perhaps 5% of the collector current. By keeping the 'on' transistor in the saturation region its effective resistance will remain low for quite high collector currents. In the example (Figure 2.2), with the full load current of 6 A flowing, the collector-emitter voltage might be say 0.33 V, giving an on-state dissipation of 2 W in the transistor when the load power is 72 W. This is not as good as a mechanical switch, but is acceptable.

We should note that the on-state base-emitter voltage is also very low, which, coupled with the small base current, means that the power required to drive the transistor is very much less than the power being switched in the collector-emitter circuit. Nevertheless, to switch the transistor in the regular pattern shown in Figure 2.2, we obviously need a base current waveform which goes on and off periodically, and we might wonder how we obtain this 'control' signal. Normally, the base drive signal will originate from a low-power oscillator (constructed from logic gates, or on a single chip), or perhaps from a microprocessor. Depending on the base circuit power requirements of the main switching transistor, it may be possible to feed it directly from the oscillator, but if necessary additional transistors are interposed between the main device and the signal source to provide the required power amplification.

Just as we have to select mechanical switches with regard to their duty, we must be careful to use the right power transistor for the job in hand. In particular, we need to ensure that when the transistor is 'on', we don't exceed the safe current, or else the active semiconductor region of the device will be destroyed by overheating. And we must make sure that the transistor is able to withstand whatever voltage appears across the collector-emitter junction when it is in the 'off' condition. If the safe voltage is exceeded, the transistor will break down, and be permanently 'on'.

A suitable heatsink will be a necessity. We have already seen that some heat is generated when the transistor is on, and at low switching rates this is the main source of unwanted heat. But at high switching rates, 'switching loss' can be very important.

Switching loss arises because the transition from on to off or vice-versa takes a finite time. We will of course arrange the base-drive circuitry so that the switching takes place as fast as possible, but in practice it will seldom take less than a few microseconds. During the switch-on period, for example, the current will be building up, while the collector-emitter voltage will be falling towards zero. The peak power reached can therefore be large, before falling to the relatively low on-state value. Of course the total energy released as heat each time the device switches is modest because the whole process happens so quickly. Hence if the switching rate is low (say once every second) the switching power loss will be insignificant in comparison with the on-state power. At high switching rates (say 100 kHz), however, when the time taken to complete the switching becomes comparable with the 'on' time, the switching power loss can easily become dominant. In practice, converters used in drives rarely employ switching rates much above 20 kHz, in order to minimise switching losses.

Chopper with inductive load – overvoltage protection

So far we have looked at chopper control of a resistive load, but in a drives context the load will usually mean the winding of a machine, which will invariably be inductive.

Chopper control of inductive loads is much the same as for resistive loads, but we have to be careful to prevent the appearance of dangerously high voltages each time the inductive load is switched 'off'. The root of the problem lies with the energy stored in the associated magnetic field. When an inductance L carries a current I, the energy stored in the magnetic field (W) is given by

$$W = \frac{1}{2}LI^2 \qquad (2.1)$$

If the inductor is supplied via a mechanical switch, and we try to open the switch with the intention of reducing the current to zero instantaneously, we are in effect seeking to destroy the stored energy. This is not possible, and what happens is that the energy is dissipated in the form of a spark across the contacts of the switch. This sparking will be familiar to anyone who has pulled off the low-voltage lead to the ignition coil in a car: serious sparking will also be observed across the contact-breaker points in the distributor whenever the spark suppression capacitor is faulty.

The appearance of a spark indicates that there is a very high voltage which is sufficient to break down the surrounding air. We can anticipate this by remembering that the voltage and current in an inductance are related by the equation

$$V = L\frac{di}{dt} \tag{2.2}$$

The voltage is proportional to the rate of change of current, so when we open the switch in order to force the current to zero quickly, a very large voltage is created in the inductance. This voltage appears across the switch, and if sufficient to break down the air, the current can continue to flow in the form of an arc.

Sparking across a mechanical switch is unlikely to cause immediate destruction, but when a transistor is used sudden death is certain unless steps are taken to tame the stored energy. The usual remedy lies in the use of a 'freewheel diode' (sometimes called a flywheel diode), as shown in Figure 2.4.

A diode is a one-way valve as far as current is concerned: it offers very little resistance to current flowing from anode to cathode, but blocks current flow from cathode to anode. Hence in the circuit of Figure 2.4(a), when the transistor is on, current (I) flows through the load, but not through the diode, which is said to be reverse-biassed (i.e. the applied voltage is trying – unsuccessfully – to push current down through the diode).

When the transistor is turned off, the current through it and the battery drops very quickly to zero. But the stored energy in

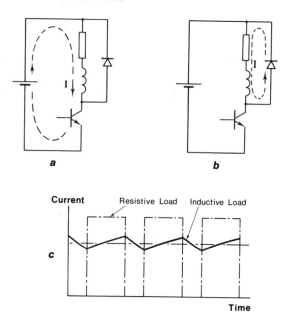

Figure 2.4 *Operation of chopper-type regulator*

the inductance means that its current cannot suddenly disappear. So since there is no longer a path through the transistor, the current diverts into the only other route available, and flows upwards through the low-resistance path offered by the diode, as shown in Figure 2.4(b).

Obviously the current no longer has a battery to drive it, so it cannot continue to flow indefinitely. In fact it will continue to 'freewheel' only until the energy originally stored in the inductance is dissipated as heat, mainly in the load resistance but also in the diode's own (low) resistance. The current waveform during chopping will then be as shown in Figure 2.4(c). Note that the current rises and falls exponentially with a time-constant of L/R (though it never reaches its steady-state value in Figure 2.4), and that the presence of inductance causes the current to be much smoother than with a purely resistive load.

Finally, we need to check that the freewheel diode prevents any dangerously high voltages from appearing across the transistor. When a diode conducts, the forward-bias volt-drop across it is small – typically 0.7 volts. Hence while the current is freewheeling, the voltage at the collector of the transistor is only 0.7 volts above the battery voltage. This 'clamping' action therefore limits the voltage across the transistor to a safe value, and allows inductive loads to be switched without damage to the switching element.

Features of power electronic converters

We can draw some important conclusions which are valid for all power electronic converters from this simple example. Firstly, efficient control of voltage (and hence power) is only feasible if a switching strategy is adopted. The load is alternately connected and disconnected from the supply by means of an electronic switch, and any average voltage up to the supply voltage can be obtained by varying the mark/space ratio. Secondly, the output voltage is not smooth d.c., but contains unwanted a.c. components which, though undesirable, are tolerable in motor drives. And finally, the load current waveform will be smoother than the voltage waveform if – as is the case with motor windings – the load is inductive.

D.C. FROM A.C. – CONTROLLED RECTIFICATION

The vast majority of drives draw their power from constant voltage 50 Hz or 60 Hz mains, and in nearly all mains converters the first stage consists of a rectifier which converts the a.c. to a crude form of d.c. Where a constant-voltage d.c. output is required, a simple (uncontrolled) diode rectifier is sufficient. But where the mean d.c. voltage has to be controllable (as in a d.c. motor drive), a controlled rectifier is used.

Many different converter configurations based on combinations of diodes and thyristors are possible, but we will focus on 'fully-controlled' converters in which all the rectifying devices are thyristors. These are especially important in the context of modern motor drives.

From the user's viewpoint, interest centres on the following questions:

- How is the output voltage controlled?
- What does the converter output voltage look like? Will there be any problems if the voltage is not pure d.c.?
- How does the range of the output voltage relate to a.c. mains voltage?

We can answer these questions without going too thoroughly into the detailed workings of the converter. This is just as well, because understanding all the ins and outs of converter operation is beyond our scope. On the other hand it is well worth trying to understand the essence of the controlled rectification process, because it assists in understanding the limitations which the converter puts on drive performance (see Chapter 4). Before tackling the questions posed above, however, it is obviously necessary to introduce the thyristor.

The thyristor

The thyristor is an electronic switch, with two main terminals (anode and cathode) and a 'switch-on' terminal (gate), as shown in Figure 2.5. Like a diode, current can only flow in the forward direction, from anode to cathode. But unlike a diode, which will conduct in the forward direction as soon as forward voltage is applied, the thyristor will continue to block forward current until a small current pulse is injected into the gate-cathode circuit, to turn it on or 'fire' it. After the gate pulse is applied, the main anode-cathode current builds up rapidly, and as soon as it reaches the 'latching' level, the gate pulse can be removed and the device will remain 'on'.

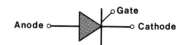

Figure 2.5   *Circuit diagram of thyristor*

Once established, the anode-cathode current cannot be interrupted by any gate signal. The non-conducting state can only be restored after the anode-cathode current has reduced to zero, and has remained at zero for the turn-off time (typically 100–200 µs in converter-grade thyristors).

When a thyristor is conducting it approximates to a closed switch, with a forward drop of only one or two volts over a wide range of current. Despite the low volt drop in the 'on' state, heat is dissipated, and heatsinks must be provided, perhaps with fan cooling. Devices must be selected with regard to the voltages to be blocked and the r.m.s. and peak currents to be carried. Their overcurrent capability is very limited, and it is usual in drives for devices to have to withstand perhaps twice full-load current for a few seconds only. Special fuses must be fitted to protect against heavy fault currents.

The reader may be wondering why we need the thyristor, since in the previous section we discussed how a transistor could be used as an electronic switch. On the face of it the transistor appears even better than the thyristor because it can be switched off while the current is flowing, whereas the thyristor will remain on until the current through it has been reduced to zero by external means. The primary reason for the use of thyristors is that they are cheaper and their voltage and current ratings extend to higher levels than power transistors. In addition, the circuit configuration in rectifiers is such that there is no need for the thyristor to be able to interrupt the flow of current, so its inability to do so is no disadvantage. Of course there are other circuits (see for example the next section dealing with inverters) where the devices need to be able to switch off on demand, in which case the transistor then has the edge over the thyristor.

Single pulse rectifier

The simplest phase-controlled rectifier circuit is shown in Figure 2.6. When the supply voltage is positive, the thyristor blocks forward current until the gate pulse arrives, and up to this point the voltage across the resistive load is zero. As soon as the device turns on the voltage across it falls to near zero, and the load voltage becomes equal to the supply voltage. When the supply voltage reaches zero, so does the current. At this point

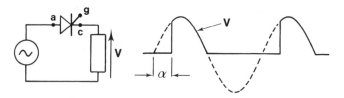

Figure 2.6   *Simple single-pulse thyristor-controlled rectifier, with resistive load and firing-angle delay α*

the thyristor regains its blocking ability, and no current flows during the negative half-cycle.

The load voltage (Figure 2.6) thus consists of parts of the positive half-cycles of the a.c. supply voltage. It is obviously not smooth, but is 'd.c.' in the sense that it has a positive mean value; and by varying the delay angle (α), measured from the zero crossing of the supply voltage, the mean voltage can be controlled.

The arrangement shown in Figure 2.6 gives only one peak in the rectified output for each complete cycle of the mains, and is therefore known as a 'single-pulse' or half-wave circuit. The output voltage (which ideally we would like to be steady d.c.) is so poor that this circuit is never used in drives. Instead, drive converters use four or six thyristors, and produce two or six pulse outputs, as will be seen in the following sections.

Single-phase fully-controlled converter – output voltage and control

The main elements of the converter circuit are shown in Figure 2.7. It comprises four thyristors, connected in bridge formation. (The term 'bridge' stems from early four-arm measuring circuits, though quite why such circuits were thought to resemble a bridge seems a mystery.)

The conventional way of drawing the circuit is shown in Figure 2.7(a), while in Figure 2.7(b) it has been redrawn to assist understanding. The top of the load can be connected (via T1) to terminal A of the mains, or (via T2) to terminal B of the mains, and likewise the bottom of the load can be connected

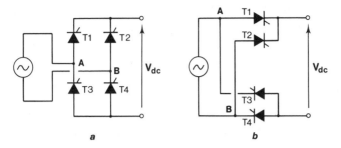

Figure 2.7   *Single-phase 2-pulse (full-wave) fully-controlled rectifier*

either to A or to B via T3 or T4 respectively.

We are naturally interested to find what the output voltage waveform on the d.c. side will look like, and in particular to discover how it can be controlled by varying the firing delay angle, α. This turns out to be more tricky than we might think, because the voltage waveform for a given α depends on the nature of the load. We will therefore look first at the case where the load is resistive, and explore the basic mechanism of phase control. Later, we will see how the converter behaves with a typical motor load.

Resistive load

Thyristors T1 and T4 are fired together when terminal A of the supply is positive, while on the other half-cycle, when B is positive, thyristors T2 and T3 are fired simultaneously. The output voltage waveform is shown by the solid line in Figure 2.8. There are two pulses per mains cycle, hence the description '2-pulse' or full-wave. At every instant the load is either

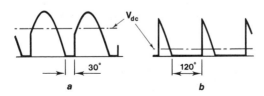

Figure 2.8   *Output voltage waveforms of single-phase fully-controlled rectifier with resistive load, for firing angle delays of 30° and 120°*

connected to the mains by the pair of switches T1 and T4, or it is connected the other way up by the pair of switches T2 and T3, or it is disconnected. The load voltage therefore consists of rectified chunks of the mains voltage. It is much smoother than in the single-pulse circuit, though again it is far from pure d.c.

The waveform in Figure 2.8(a) corresponds to $\alpha = 30°$, while Figure 2.8(b) is for $\alpha = 120°$. The mean value, V_{dc} is shown in each case. It is clear that the larger the delay angle, the lower the output voltage. The maximum output voltage (V_{do}) is obtained with $\alpha = 0°$, and is given by

$$V_{do} = \frac{2}{\pi}\sqrt{2}V_{rms} \qquad (2.3)$$

where V_{rms} is the r.m.s. voltage of the incoming mains. We note that when α is zero, the output voltage is the same as it would be for an uncontrolled diode bridge rectifier, since the thyristors conduct for the whole of the half-cycle for which they are forward-biassed. The variation of the mean d.c. voltage with α is given by

$$V_{dc} = \frac{1}{2}(1 + \cos\alpha)\,V_{do} \qquad (2.4)$$

from which we see that with a resistive load the d.c. voltage can be varied from a maximum of V_{do} down to zero by varying α from 0° to 180°.

Inductive (motor) load

As mentioned above, motor loads are inductive, and we have seen earlier that the current cannot change instantaneously in an inductive load. We must therefore expect the behaviour of the converter with an inductive load to differ from that with a resistive load, in which the current was seen to change instantaneously.

At first sight the fact that the behaviour of the converter depends on the nature of the load is a most unwelcome prospect. What we would like is to be able to say that, regardless of the load, we can specify the output voltage waveform once we have fixed the delay angle α. We would then know what value of α to select to achieve any desired mean output voltage. What we find in practice is that once we have fixed α, the mean output voltage with a resistive-inductive load is not the same as with a purely resistive load, and therefore we cannot give a simple general formula for the mean output voltage in terms of α.

Fortunately, however, it turns out that the output voltage waveform for a given α does become independent of the load inductance once there is sufficient inductance to prevent the load current from ever falling to zero. This condition is known as 'continuous current', and, happily, many motor circuits do have sufficient self-inductance to ensure that we achieve continuous current. Under continuous current conditions, the output voltage waveform only depends on the firing angle, and not on the actual inductance present. This makes things much more straightforward, and typical output voltage waveforms for this continuous current condition are shown in Figure 2.9.

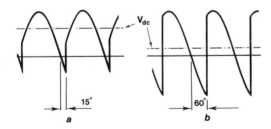

Figure 2.9 *Output voltage waveforms of single-phase fully-controlled rectifier supplying an inductive (motor) load, for firing angle delays of 15° and 60°*

The waveform in Figure 2.9(a) corresponds to α = 15°, while Figure 2.9(b) corresponds to α = 60°. We see that, as with the resistive load, the larger the delay angle the lower the mean output voltage. However with the resistive load the output voltage was never negative, whereas we see that for short periods the output voltage can now become negative. This is because the inductance smooths out the current (see Figure 4.2, for example) so that at no time does it fall to zero. As a result, one or other pair of thyristors is always conducting, and there is no time at which the load is disconnected from the a.c. supply.

The maximum voltage (V_{do}) is again obtained when α is zero, and is the same as for the resistive load (equation 2.3). It is easy to show that the mean d.c. voltage is now related to α by

$$V_{dc} = V_{do}\cos\alpha \qquad (2.5)$$

This equation indicates that we can control the mean output voltage by controlling α, though equation 2.3 shows that the variation of mean voltage with α is different from that for a resistive load (equation 2.4). We also see that when α is greater than 90° the mean output voltage is negative. The fact that we can obtain a nett negative output voltage with an inductive load contrasts sharply with the resistive load case, where the output voltage could never be negative. We shall see later that this facility allows the converter to return energy from the load to the supply, and this is important when we want to use the converter with a d.c. motor in the regenerating mode.

3-phase fully-controlled converter

The main power elements are shown in Figure 2.10. The three-phase bridge has only two more thyristors than the single-phase bridge, but the output voltage waveform is very much better, as shown in Figure 2.11. There are now six pulses of the output voltage per mains cycle, hence the description '6-pulse'. The thyristors are again fired in pairs (one in the top half of the bridge and one – from a different leg – in the bottom half), and each thyristor carries the output current for one third of the time. As in the single-phase converter, the delay angle controls the output voltage, but now $\alpha = 0°$ corresponds to the point at which the phase voltages are equal (see Figure 2.11).

The enormous improvement in the smoothness of the output voltage waveform is clear when we compare Figures 2.11 and 2.9, and it indicates the wisdom of choosing a 3-phase converter whenever possible. The much better voltage waveform also means that the desirable 'continuous current' condition is

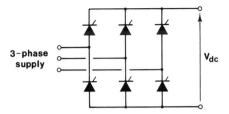

3-phase
supply

V_{dc}

Figure 2.10 *Three-phase fully-controlled thyristor converter*

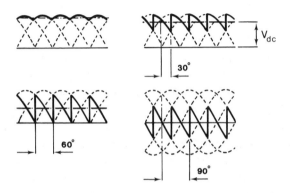

V_{dc}

30°

60°

90°

Figure 2.11 *Output voltage waveforms for three-phase fully-controlled converter supplying an inductive (motor) load, with various firing delay angles from $\alpha = 0°$ to $90°$. The mean d.c. voltage is shown by the horizontal dotted line, except for $\alpha = 90°$ where the mean d.c. voltage is zero*

much more likely to be met, and the waveforms in Figure 2.11 have therefore been drawn with the assumption that the load current is in fact continuous. Occasionally, even a six-pulse waveform is not sufficiently smooth, and some very large drive converters therefore consist of two six-pulse converters with their outputs in series. A phase-shifting transformer is used to insert a 30° shift between the a.c. supplies to the two 3-phase bridges. The resultant ripple voltage is then 12-pulse.

Returning to the 6-pulse converter, the mean output voltage can be shown to be given by

$$V_{dc} = V_{do}\cos\alpha = \frac{3}{\pi}\sqrt{2}V_{rms}\cos\alpha \qquad (2.6)$$

We note that we can obtain the full range of output voltages from $+ V_{do}$ to $- V_{do}$, so that, as with the single-phase converter, regenerative operation will be possible.

Output voltage range

In Chapter 4 we will discuss the use of the fully-controlled converter to drive a d.c. motor, so it is appropriate at this stage to look briefly at the typical voltages we can expect. Mains a.c. supply voltages obviously vary around the world, but single-phase supplies are usually 220–240 V, and we see from equation 2.3 that the maximum mean d.c. voltage available from a single phase 240 V supply is 216 V. This is suitable for 180–200 V motors. If a higher voltage is needed (for say a 300 V motor), a transformer must be used to step up the mains.

Turning now to typical three-phase supplies, the lowest three-phase industrial voltages are usually around 380–440 V. (Higher voltages of up to 11 kV are used for large drives, but these will not be discussed here.) So with $V_{rms} = 415$ V for example, the maximum d.c. output voltage (equation 2.6) is 560 volts. After allowances have been made for supply variations and impedance drops, we could not rely on obtaining much more than 520–540 V, and it is usual for the motors used with 6-pulse drives fed from 415 V, 3-phase supplies to be rated in the range 440–500 V. (Often the motor's field winding will be supplied from single phase 240 V, and field voltage ratings are then around 180–200 V, to allow a margin in hand from the theoretical maximum of 216 V referred to earlier.)

Firing circuits

Since the gate pulses are only of low power, the gate drive circuitry is simple and cheap. Often a single integrated circuit (chip) contains all the circuitry for generating the gate pulses, and for synchronising them with the appropriate delay angle α with respect to the supply voltage. To avoid direct electrical connection between the high voltages in the main power circuit and the low voltages used in the control circuits, the gate pulses are usually coupled to the thyristor by means of small pulse transformers. Most converters include inverse cosine-weighted firing circuitry, so that the mean output voltage of the converter is directly proportional to the input control voltage, which typically ranges from 0 to 10 volts.

A.C. FROM D.C. – INVERSION

The business of getting a.c. from d.c. is known as inversion, and nine times out of ten we would ideally like to be able to produce sinusoidal voltages of whatever frequency and amplitude we choose. Unfortunately the constraints imposed by the necessity to use a switching strategy mean that we always have to settle for a voltage waveform which is composed of rectangular chunks, and is thus far from ideal. Nevertheless it turns out that a.c. motors are remarkably tolerant, and are happy to operate despite the inferior waveforms produced by the inverter.

Single-phase inverter

We can illustrate the basis of inverter operation by considering the single-phase example shown in Figure 2.12.

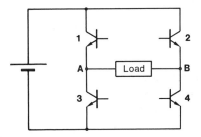

Figure 2.12   *Inverter circuit for single-phase output*

The input or d.c. side of the inverter is usually referred to as the 'd.c. link', reflecting the fact that in the majority of cases the d.c. is obtained by rectifying the incoming constant-frequency mains. The output or a.c. side is taken from terminals A and B in Figure 2.12.

When transistors 1 and 4 are switched on, the load voltage is positive, and equal to the d.c. link voltage, while when 2 and 3 are on it is negative. If no devices are switched on, the output voltage is zero. Typical output voltage waveforms at low and high switching frequencies are shown in Figures 2.13(a) and 2.13(b) respectively.

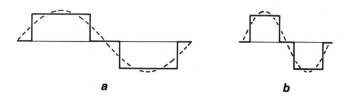

Figure 2.13   *Inverter output voltage waveforms – resistive load*

Here each pair of devices is on for one-third of a cycle, and all the devices are off for two periods of one-sixth of a cycle. The output waveform is clearly not a sinewave, but at least it is alternating and symmetrical. The fundamental component is shown dotted in Figure 2.13.

Within each cycle the pattern of switching is regular, and easily programmed using appropriate logic circuitry. Frequency variation is obtained by altering the clock frequency controlling the 4-step switching pattern. (The oscillator providing the clock signal can be controlled by an analogue voltage, or it can be generated in software.) The effect of varying the switching frequency is shown in Figure 2.13, from which we can see that the amplitude of the fundamental component of voltage remains constant, regardless of frequency. Unfortunately (as explained in Chapter 7) this is not what we want for supplying an induction motor: to prevent the air-gap flux in the motor from falling as the frequency is raised we need to be able to increase the voltage in proportion to the frequency. We will look at voltage control shortly, after a brief digression to discuss the problem of 'shoot-through'.

Inverters with the configurations shown in Figures 2.12 and 2.15 are subject to a potentially damaging condition which can arise if both transistors in one 'leg' of the inverter inadvertently turn on simultaneously. This should never happen if the devices are switched correctly, but if something goes wrong and both devices are on together – even for a very short time – they form a short-circuit across the d.c. link. This fault condition is referred to as 'shoot-through' because a high current is established very rapidly, destroying the devices. A good inverter therefore includes provision for protecting against the possibility of shoot-through, usually by imposing a minimum time-delay between one device in the leg going off and the other coming on.

Output voltage control

There are two ways in which the amplitude of the output voltage can be controlled. First, if the d.c. link is provided from a.c. mains via a controlled rectifier or from a battery via a chopper, the d.c. link voltage can be varied. We can then set the amplitude of the output voltage to any value within the range of the link. For a.c. motor drives (see Chapter 7) we can arrange for the link voltage to track the output frequency of the inverter, so that at high output frequency we obtain a high output voltage and vice-versa. This method of voltage control results in a simple inverter, but requires a controlled (and thus relatively expensive) rectifier for the d.c. link.

The second method, which now predominates in small and medium sizes, achieves voltage control by pulse width modulation (PWM) within the inverter itself. A cheaper uncontrolled rectifier can then be used to provide a constant-voltage d.c. link.

The principle of voltage control by PWM is illustrated in Figure 2.14.

At low output frequencies, a low output voltage is usually required, so one of each pair of devices is used to chop the voltage, the mark–space ratio being varied to achieve the desired voltage at the output. The low fundamental voltage component at low frequency is shown dotted in Figure 2.14(a). At a higher frequency a higher voltage is needed, so the chopping device is allowed to conduct for a longer fraction of each cycle, giving the higher fundamental output shown in Figure 2.14(b). As the frequency is raised still higher, the

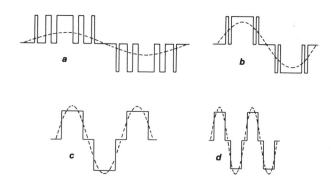

Figure 2.14 *Inverter output voltage and frequency control by pulse-width modulation*

separate 'on' periods eventually merge, giving the waveform shown in Figure 2.14(c). Any further increase in frequency takes place without further increase in the output voltage, as shown in Figure 2.14(d).

In drives applications, the range of frequencies over which the voltage/frequency ratio can be kept constant is known as the 'PWM' region, and the upper limit of the range is usually taken to define the 'base speed' of the motor. Above this frequency, the inverter can no longer match voltage to frequency, the inverter effectively having 'run out of steam' as far as voltage is concerned. The maximum voltage is thus governed by the link voltage, which must therefore be sufficiently high to provide whatever fundamental voltage the motor needs at its base speed, which is usually 50 or 60 Hz.

Beyond the PWM region the voltage waveform is as shown in Figure 2.14(d): this waveform is invariably referred to as 'quasi-square', though in the light of the overall object of the exercise (to approximate to a sinewave) a more accurate description would be 'quasi-sine'.

When supplying an inductive motor load, fast recovery freewheel diodes are needed in parallel with each device. These may be discrete devices, or fitted in a common package with the transistor, or even integrated to form a single transistor/diode device.

Sinusoidal PWM

So far we have emphasised the importance of being able to control the amplitude of the fundamental output voltage by modulating the width of the pulses which make up the output waveform. If this were the only requirement, we would have an infinite range of modulation patterns which would do. But as well as the right fundamental amplitude, we want the harmonic content to be minimised, i.e. we want the output waveform to be as close as possible to a pure sinewave. It is particularly important to limit the amplitude of the low-order harmonics, since these are the ones which are most likely to provoke an unwanted response from the motor.

The number, width, and spacing of the pulses is therefore optimised to keep the harmonic content as low as possible. A host of sophisticated strategies have been developed, almost all using a microprocessor based system to store and/or generate the modulation patterns. There is an obvious advantage in using a high switching frequency, since there are then more pulses to play with. Ultrasonic frequencies are now widely used, and as devices improve the switching frequencies may go still higher. Most manufacturers claim that their particular system is better than the competition, but it is too early to say which will ultimately emerge as best for motor operation. Some early schemes used comparatively few pulses per cycle, and changed the number of pulses in discrete steps rather than smoothly, which earned them the nickname 'gear-changers'. These inverters were noisy and irritating.

3-phase inverter

A 3-phase output can be obtained by adding only two more switches to the four needed for a single-phase inverter, giving the typical power-circuit configuration shown in Figure 2.15. A freewheel diode is required in parallel with each transistor to protect against overvoltages caused by an inductive (motor) load.

We note that the circuit configuration in Figure 2.15 is the same as for the 3-phase controlled rectifier looked at earlier.

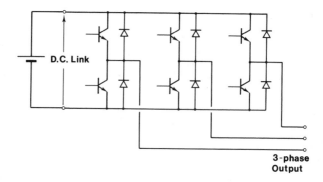

Figure 2.15   *Three-phase inverter power circuit*

We mentioned then that the controlled rectifier could be used to regenerate, i.e. to convert power from d.c. to a.c., and this is of course 'inversion' as we now understand it.

Forced and natural commutation

We have assumed in this discussion that the switching devices can turn off or 'commutate' on demand, so that the output (load) current is either reduced to zero or directed to another leg of the inverter. Transistors and gate-turn-off thyristors (see next section) can operate like this, but, as explained earlier, conventional thyristors cannot turn off on command. Nevertheless thyristors are widely used to invert power from d.c. to a.c., as we will see when we look at the d.c. motor drive in Chapter 4, so we must be clear how this is possible.

There are two distinct ways in which thyristors are used in inverters. In the first, where the inverter is used to supply an essentially passive load, such as an induction motor, each thyristor has to be equipped with its own auxiliary 'forced commutating' circuit, whose task is to force the current through the thyristor to zero when the time comes for it to turn off. The commutating circuits are quite complex, and require substantial capacitors to store the energy required for commutation. Force commutated thyristor converters therefore tend to be bulky and expensive. At one time they were the only alternative, but they are now more or less obsolete, having been superseded by transistor or GTO versions.

The other way in which thyristors can be used to invert power from d.c. to a.c. is when the a.c. side of the bridge is connected to a 3-phase mains supply. This is the normal 'controlled rectifier' arrangement introduced earlier. In this case it turns out that the currents in the thyristors are naturally forced to zero by the active mains voltages, thereby allowing the thyristors to commutate or turn off naturally. This mode of operation continues to be important, as we will see when we look at d.c. motor drives.

INVERTER SWITCHING DEVICES

As far as the user is concerned, it does not really matter what type of switching device is used inside the inverter, but it is probably helpful to mention the four most important families of devices in current use so that the terminology is familiar and the symbols used for each device can be recognised. The feature which unites all four devices is that they can be switched on and off by means of a low-power control signal, i.e. they are self-commutating. We have seen earlier that this ability to be turned on or off on demand is essential in any inverter which feeds a passive load, such as an induction motor.

Each device is discussed briefly below, with a broad indication of its most likely range of application. Because there is considerable overlap between competing devices, it is not possible to be dogmatic and specify which device is best, and the reader should not be surprised to find that one manufacturer may offer a 5 kW (6.7 h.p.) inverter which uses MOSFETs while another chooses to use IGBTs. The whole business of power electronics is really still in its infancy, and there are doubtless other devices yet to emerge. One trend which does look certain to continue is the move to integrate the drive and protection circuitry in the same package as the switching device (or devices). This obviously leads to considerable simplification and economy in the construction of the complete converter.

Bipolar junction transistor (BJT)

Historically the bipolar junction transistor was the first to be used for power switching. Of the two versions (npn and pnp) only the npn has been widely used in inverters for drives, mainly in applications ranging up to a few kW and several hundred volts.

The npn version is shown in Figure 2.16: the main (load) current flows into the collector (C) and out of the emitter (E), as shown by the arrow on the device symbol. To switch the device on (i.e. to make the resistance of the collector-emitter circuit low, so that load current can flow) a small current must be caused to flow from the base (B) to the emitter. When the base-emitter current is zero, the resistance of the collector-emitter circuit is very high, and the device is switched off.

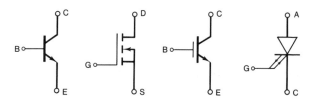

Figure 2.16   *Circuit symbols for self-commutating devices, showing (from L to R), BJT, MOSFET, IGBT, GTO*

The advantage of the bipolar transistor is that when it is turned on, the collector-emitter voltage is low (see Figure 2.3) and hence the power dissipation is small in comparison with the load power, i.e. the device is an efficient power switch. The disadvantage is that although the power required in the base-emitter circuit is tiny in comparison with the load power, it is not insignificant and in the largest power transistors can amount to several tens of watts. This means that the complexity and cost of the base drive circuitry can be considerable.

Metal oxide semiconductor field effect transistor (MOSFET)

Since the 1980s the power MOSFET has gradually superseded the BJT in inverters for drives. Like the BJT, the MOSFET is a three-terminal device and is available in two versions, the n-channel and the p-channel. The n-channel is the most widely used, and is shown in Figure 2.16. The main (load) current flows into the drain (D) and out of the source (S). (Confusingly, the load current in this case flows in the *opposite* direction to the arrow on the symbol.) Unlike the BJT, which is controlled by the base current, the MOSFET is controlled by the gate-source voltage.

To turn the device on, the gate-source voltage must be comfortably above a threshold (typically between 2 and 4 volts). When the voltage is first applied to the gate, currents flow in the parasitic gate-source and gate-drain capacitances, but once these capacitances have been charged the input current to the gate is negligible, so the steady-state gate drive power is negligible. To turn the device off, the parasitic capacitances must be discharged and the gate-source voltage must be held below the threshold level.

The principal advantage of the MOSFET is that it is a voltage-controlled device which requires negligible power to hold it in the 'on' state. The gate drive circuitry is thus less complex and costly than the base drive circuitry of an equivalent bipolar device. The disadvantage of the MOSFET is that in the 'on' state the effective resistance of the drain-source is higher than an equivalent bipolar device, so the power dissipation is higher and the device is rather less efficient as a power switch. MOSFETs are used in low and medium power inverters up to a few kW, with voltages generally not exceeding 700 V.

Insulated gate bipolar transistor (IGBT)

The IGBT (Figure 2.16) is a hybrid device which combines the best features of the MOSFET (i.e. ease of gate turn-on and turn-off from low-power logic circuits) and the BJT (relatively low power dissipation in the main collector-emitter circuit). These obvious advantages give the IGBT the edge over the MOSFET and BJT, and account for the widespread take-up of the new technology amongst inverter drive manufacturers in the early 1990s. They are particularly well suited to the medium power, medium voltage range (up to several hundred kW or h.p.).

The path for the main (load) current is from collector to emitter, as in the npn bipolar device.

Gate turn-off thyristor (GTO)

The GTO (Figure 2.16) is turned on by a pulse of current in the gate-cathode circuit in much the same way as a conventional thyristor. But unlike an ordinary thyristor, which cannot be turned off by gate action, the GTO can be turned off by a negative gate-cathode current. The main (load) current flows from anode to cathode, as in a conventional thyristor. The twin arrowed paths on the gate lead (Figure 2.16) indicate that control action is achieved by both forward and reverse gate currents. (In US literature, a single gate lead with a short crossbar is used instead of the two arrows.)

The gate drive requirements are more demanding than for a conventional thyristor, and the on-state performance is worse, with a forward volt-drop of perhaps 3 V compared with 1.5 V, but these are the penalties to be paid in return for the added flexibility. The GTO has considerably higher voltage and current ratings (up to 3 kV and 2 kA) than the other three devices and is therefore used in high power inverters.

CONVERTER WAVEFORMS AND ACOUSTIC NOISE

In common with most textbooks, the waveforms shown in this chapter (and later in the book) are what we would hope to see under ideal conditions. It makes sense to concentrate on these ideal waveforms from the point of view of gaining a basic understanding, but we ought to be warned that what we see on an oscilloscope may well look rather different!

We have seen that the essence of power electronics is the switching process, so it should not come as much of a surprise to learn that in practice the switching is seldom achieved in such a clear-cut fashion as we have assumed. Usually, there will be some sort of high-frequency 'ringing' evident, particularly

Plate 2.1 *Range of d.c. and a.c. motor drive converters ranging from 500 W to 50 kW (0.67 to 67 h.p.). The low-power control circuitry is visible at the front, the power switching devices being mounted on heatsinks at the rear. The heatsinks and cooling fans are visible on the larger units. (Photograph by courtesy of CEGELEC Industrial Controls Ltd)*

on the voltage waveforms following each transition due to switching. This is due to the effects of stray capacitance and inductance: it will have been anticipated at the design stage, and steps will have been taken to minimise it by fitting 'snubbing' circuits at the appropriate places in the converter. However complete suppression of all these transient phenomena is seldom economically worthwhile so the user should not be too alarmed to see remnants of the transient phenomena in the output waveforms.

Acoustic noise is also a matter which can worry newcomers. Most power electronic converters emit whining or humming sounds at frequencies corresponding to the fundamental and harmonics of the switching frequency, though when the converter is used to feed a motor, the sound from the motor is usually a good deal louder than the sound from the converter itself. These sounds are very difficult to describe in words, but typically range from a high-pitched hum through a whine to a piercing whistle. They vary in intensity with the size of converter and the load, and to the trained ear can give a good indication of the health of the motor and converter.

COOLING OF POWER SWITCHING DEVICES

Thermal resistance

We have seen that by adopting a switching strategy the power loss in the switching devices is small in comparison with the power throughput, so the converter has a high efficiency. Nevertheless almost all the heat which is produced in the switching devices is released in the active region of the semiconductor, which is itself very small and will overheat and break down unless it is adequately cooled. It is therefore essential to ensure that even under the most onerous operating conditions, the temperature of the active junction inside the device does not exceed the safe value.

Consider what happens to the temperature of the junction region of the device when we start from cold (i.e. ambient) temperature and operate the device so that its average power dissipation remains constant. At first, the junction temperature begins to rise, so some of the heat generated is conducted to the metal case, which stores some heat as its temperature rises. Heat then flows into the heatsink (if fitted) which begins to warm up, and heat begins to flow to the surrounding air, at ambient temperature. The temperatures of the junction, case and heatsink continue to rise until eventually an equilibrium is reached when the total rate of loss of heat to ambient temperature is equal to the power dissipation inside the device.

The final steady-state junction temperature thus depends on how difficult it is for the power loss to escape down the temperature gradient to ambient, or in other words on the total 'thermal resistance' between the junction inside the device and the surrounding medium (usually air). Thermal resistance is usually expressed in °C/watt, which directly indicates how much temperature rise will occur in the steady state for every watt of dissipated power. It follows that for a given power dissipation, the higher the thermal resistance, the higher the temperature rise, so in order to minimise the temperature rise of the device, the total thermal resistance between it and the surrounding air must be made as small as possible.

The device designer aims to minimise the thermal resistance between the semiconductor junction and the case of the device, and provides a large and flat metal mounting surface to minimise the thermal resistance between the case and the heatsink. The converter designer must ensure good thermal contact between the device and the heatsink, usually by a bolted joint liberally smeared with heat-conducting compound to fill any microscopic voids, and must design the heatsink to minimise the thermal resistance to air (or in some cases oil or water). Heatsink design offers the only real scope for appreciably reducing the total resistance, and involves careful selection of the material, size, shape and orientation of the heatsink, and the associated air-moving system (see below).

One drawback of the good thermal path between the junction and case of the device is that the metal mounting surface (or surfaces in the case of the popular hockey-puck package) are often electrically 'live'. This poses a difficulty for the converter designer, because mounting the device directly on the heatsink causes the latter to be electrically live and therefore hazardous. In addition, several separate isolated heatsinks may be required in order to avoid short-circuits. The alternative is for the devices to be electrically isolated from the heatsink using thin mica spacers, but then the thermal resistance is appreciably increased.

The increasing trend towards packaged 'modules' with an electrically isolated metal base gets round this problem. These contain combinations of transistors, diodes or thyristors, from which various converter circuits can be built up. Several modules can be mounted on a single heatsink, which does not have to be isolated from the enclosure or cabinet. Currently they are available in ratings suitable for converters up to many tens of kW or h.p., and the range is expanding. This development, coupled with a move to fan-assisted cooling of heatsinks, has resulted in a dramatic reduction in the overall size of complete converters, so that a modern 20 kW (26.8 h.p.) drive converter is perhaps only the size of a small briefcase.

Arrangement of heatsinks and forced-air cooling

The principal factors which govern the thermal resistance of a heatsink are the total surface area, the condition of the surface and the air flow. Most converters use extruded aluminium heatsinks, with multiple fins to increase the effective cooling surface area and lower the resistance, and with a machined face or faces for mounting the devices. Heatsinks are usually mounted vertically to improve natural air convection. Surface finish is important, with black anodised aluminium being typically 30% better than bright. At one time heatsinks could only be bought from the manufacturer, but a wide range of compact and highly efficient heatsinks and fans designed specifically for cooling electronic equipment is now readily available from the leading electronic component suppliers.

A typical layout for a medium-power (say 100 kW or 134 h.p.) converter is shown in Figure 2.17.

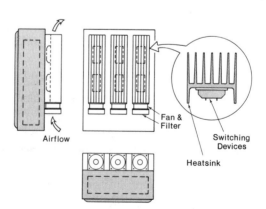

Figure 2.17   *Typical layout of converter showing heatsinks and cooling fans*

The fan(s) are positioned either at the top or bottom of the heatsink, and draw external air upwards, assisting natural convection. The value of even a modest air-flow is shown by the sketch in Figure 2.18. With an air velocity of only 2 m/sec, for example, the thermal resistance is halved as compared with the naturally-cooled set-up, which means that for a given temperature rise the heatsink can be half the size of the naturally-cooled one. Only a little of this saving in space is taken up by the fans, as shown Figure 2.17. Large increases in the air velocity bring diminishing returns, as shown in Figure 2.18, and also introduce additional noise which is generally undesirable.

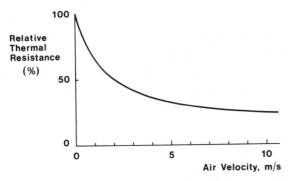

Figure 2.18   *Sketch indicating influence of air velocity on effective thermal resistance. (The thermal resistance in still air is taken as 100%)*

Cooling fans

Cooling fans have integral hub-mounted inside-out motors, i.e. the rotor is outside the stator and carries the blades of the fan. The rotor diameter/length ratio is much higher than for most conventional motors in order to give a slimline shape to the fan assembly, which is well-suited for mounting at the end of an extruded heatsink (Figure 2.17). The rotor inertia is thus relatively high, but this is unimportant because the total inertia is dominated by the impeller, and there is no need for high accelerations.

Mains voltage 50 or 60 Hz fans have external-rotor single-phase shaded-pole motors, which normally run at a fixed speed of around 2700 rev/min, and have input powers typically between 10 and 50 W (0.01–0.05 h.p.). The torque required in a fan is roughly proportional to the cube of the speed, so the starting torque requirement is low and the motor can be designed to have a high running efficiency (see Chapter 6). Where acoustic noise is a problem slower-speed (but less efficient) versions are used.

Low-voltage (5, 12 or 24 V) d.c. fans employ brushless motors with Hall-effect rotor position detection (see Chapter 9). The absence of sparking from a conventional commutator is important to limit interference with adjacent sensitive equipment. These fans are generally of lower power than their a.c. counterparts, typically from as little as 1 W up to about 10 W, and with running speeds of typically between 3000 and 5000 rev/min. They are mainly used for cooling circuit boards directly, and have the advantage that the speed can be controlled by voltage variation, thereby permitting a trade-off between noise and volume flow.

QUESTIONS ARISING

- *Why is it rare in drives to find power electronic switching converters operating at much more than a few tens of kHz; wouldn't it be possible to achieve a closer approximation to the desired output waveshape by using a higher switching rate?*

The waveforms could indeed be improved, but the problem is that every time a device switches there is a finite time during which neither the voltage across it nor the current through it is zero. During this 'switching time' the power dissipation is very high. The average switching dissipation in the device is therefore proportional to the frequency, so if the frequency is too high, the device will overheat and burn out.

- *I can see that the d.c. voltage from a 6-pulse converter is much smoother than the voltage from a 2-pulse converter, but why not smooth all such waveforms by connecting a capacitor across the output? This seems to be standard practice in the bench-type electronic d.c. power supplies I have seen.*

The simple answer is that it is not necessary for the voltage to be smooth in a d.c. motor drive: it is much more important for the current to be smooth (see Chapter 4), and in this respect the motor armature inductance is very beneficial. This is fortunate, because the power levels for the majority of drives are such that in order to store enough energy to smooth out the rectified voltage, very bulky (and expensive) capacitors would be required. (In contrast, the voltage must be smooth in an electronic power supply, and because the power levels are low, it is feasible to use capacitors to achieve the initial smoothing, and to follow this by a linear or chopper-type regulator.)

- *Mercury arc rectifiers are still to be seen driving d.c. motors. Are they similar to thyristor converters, and would it be possible to replace an ageing mercury rectifier with a thyristor unit?*

Yes, a mercury arc rectifier behaves in a similar way to a solid-state diode. The MAR has an anode in the vapour above a cathode pool of mercury: the vapour will conduct current from anode to cathode, but not vice-versa. The grid-controlled MAR is similar but has a fine grid between anode and cathode. By varying the voltage on the grid the anode-cathode voltage at which the valve begins to conduct can be controlled. Hence the MAR can be used as a controlled rectifier, the grid voltage being adjusted to control the point in the cycle when conduction starts, in much the same way as conduction is initiated in the thyristor by 'firing' it.

The bridge connection (see Figure 2.10) was not often used for MARs because it was more economic to make a multi-anode rectifier with a single pool of mercury, so all the cathodes had to be common. Six-pulse drives rectifiers required a six-phase supply which necessitated a three-phase transformer with six separate secondary windings.

Replacement with a thyristor converter (complete with controls) makes sense if the motor is still in good condition. Depending on the motor voltage rating, it may be possible to use a thyristor bridge directly from the original mains supply, but in some cases a new transformer will be required.

- *I overheard someone say that a d.c. chopper drive is like a transformer. What did they mean?*

An ideal d.c. to d.c. chopper is 100% efficient, so the average power input is equal to the average power output. We can therefore view the chopper as a means of providing a low voltage, high current output from a high voltage, low current input (or vice-versa), which is precisely what we expect in a conventional a.c. transformer.

To illustrate, suppose that we were using a step-down chopper fed from a 100 V battery to supply a 20 volt motor drawing a mean current of 10 A. The average motor voltage (20 V) results from an 'on' pulse of 100 V for time τ, followed by a zero-voltage 'freewheeling' period of 4τ during each chopping cycle. Assuming that there is sufficient motor inductance to make the current smooth, the current is constant at 10 A. The average power in the motor is thus $20 \times 10 = 200$ W.

The battery voltage is constant at 100 V, and the battery current consists of pulses of 10 A for time τ, followed by zero current for time 4τ. The average battery current is thus 2 A, and the battery power is $100 \times 2 = 200$ W, which is equal to the motor power.

- *Is it true that one can get a nasty shock from the d.c. side of a controlled rectifier even when the d.c. output voltage is set at a very low value?*

Yes. Whatever the mean d.c. voltage, the output terminals will always be live with respect to earth, since a pair of thyristors (see Figure 2.10) will always be conducting. There will also be a large a.c. voltage across the output, even when the d.c. voltage is zero, as shown by the sketch for $\alpha = 90°$ in Figure 2.11. In this condition a voltmeter on the d.c. range will read zero, but on the a.c. range it may well read several hundred volts, depending on the incoming mains voltage.

- *Why is it that hi-fi amplifiers can produce beautiful sinewaves, but power electronic converters always produce crude approximations, such as those shown in Fig 2.14?*

Power electronic converters must be efficient, which means that the solid-state devices used to regulate power flow can only be used as switches. In both the on and off states, the power loss in the device is small, so it does not overheat and burn out. But this means that we can only approximate the desired output waveforms. In addition, it is not usually practicable to filter the waveforms because the filter components are too bulky and expensive.

None of the arguments above apply with the same force at low power levels, and it is even possible to operate the devices (e.g. the transistor) in the 'linear' region, i.e. as a variable resistor. The power dissipation in the device is much higher than when switching the same load, but since the absolute powers are lower, the device can survive.

3

CONVENTIONAL D.C. MOTORS

INTRODUCTION

Until the 1980s the conventional (brushed) d.c. machine was the automatic choice where speed or torque control is called for, and it retains a large share of the market despite the growing challenge from the inverter-fed induction motor. Applications range from steel rolling mills, railway traction, through a very wide range of industrial drives to robotics, printers and precision servos. The range of power outputs is correspondingly wide, from several megawatts at the top end down to a only a few watts, but except for a few of the small low-performance ones, such as those used in toys, all have the same basic structure, as shown in Figure 3.1.

As in any electrical machine it is possible to design a d.c. motor for any desired supply voltage, but for several reasons it is unusual to find rated voltages lower than about 6 V or much higher than 700 V. The lower limit arises because the brushes (see below) give rise to an unavoidable volt-drop of perhaps 0.5–1 V, and it is clearly not good practice to let this 'wasted' voltage became a large fraction of the supply voltage. At the other end of the scale it becomes prohibitively expensive to insulate the commutator segments to withstand higher voltages. The function and operation of the commutator are discussed later, but it is appropriate to mention here that brushes and commutators are troublesome at very high speeds.

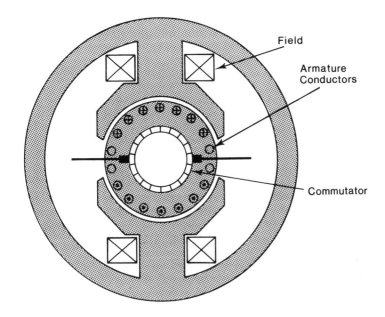

Figure 3.1 *Conventional (brushed) d.c. motor*

Small d.c. motors, say up to hundreds of watts output, can run at perhaps 12 000 rev/min, but the majority of medium and large motors are usually designed for speeds below 3000 rev/min.

Increasingly, motors are being supplied with power-electronic drives, which draw power from the a.c. mains and convert it to d.c. for the motor. Since the mains voltages tend to be standardised (e.g. 110 V, 220–240 V, or 380–440 V, 50 or 60 Hz), motors are made with rated voltages which match the range of d.c. outputs from the converter (see Chapter 2).

As mentioned above, it is quite normal for a motor of a given power, speed and size to be available in a range of different voltages. In principle all that has to be done is to alter the number of turns and the size of wire making up the coils in the machine. A 12 V, 4 A motor, for example, could easily be made to operate from 24 V instead, by winding its coils with twice as many turns of wire having only half the cross-sectional area of the original. The full speed would be the same at 24 V as the original was at 12 V, and the rated current would be 2 A, rather than 4 A. The input power and output power would be unchanged, and externally there would be no change in appearance, except that the terminals might be a bit smaller.

Traditionally d.c. motors were classified as shunt, series, or separately excited. In addition it was common to see motors referred to as 'compound-wound.' These descriptions date from the period before the advent of power electronics, and a strong association built up linking one or other 'type' of d.c. machine with a particular application. There is really no fundamental difference between shunt, series or separately excited machines, and the names simply reflect the way in which the field and armature circuits are interconnected. The terms still persist, however, and we will refer to them again later. But first we must gain an understanding of how the basic machine operates, so that we are equipped to understand what the various historic terms mean, and hence see how modern practice is deployed to achieve the same ends.

We should make clear at this point that whereas in an a.c. machine the number of poles is of prime importance in determining the speed, the pole-number in a d.c. machine is of little consequence as far as the user is concerned. It turns out to be more economical to use two or four poles in small or medium size d.c. motors, and more (e.g. ten or twelve or even more) in large ones, but the only difference to the user is that the 2-pole type will have two brushes at 180°, the 4-pole will have four brushes at 90°, and so on. Most of our discussion centres on the 2-pole version in the interests of simplicity, but there is no essential difference as far as operating characteristics are concerned.

TORQUE PRODUCTION

Torque is produced by interaction between the axial current-carrying conductors on the rotor and the radial magnetic flux produced by the stator. This flux or 'excitation' can be furnished by permanent magnets (Figure 3.2(a)) or by means of field windings (Figure 3.2(b)).

Plate 3.1   *Permanent-magnet d.c. motor, 500 W (0.67 h.p.) at 2000 rev/min. (Photograph by courtesy of Brook Crompton)*

Permanent magnet versions are available in motors with outputs from a few watts up to a few kilowatts, while wound-field machines begin at about 100 watts and extend to

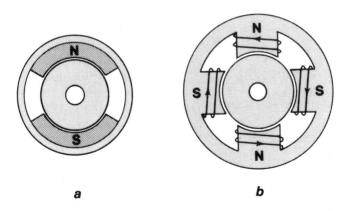

a	*b*

Figure 3.2   *Excitation (field) systems for d.c. motors (a) 2-pole permanent magnet; (b) 4-pole wound field*

the largest (MW) outputs. The advantages of the permanent magnet type are that no electrical supply is required for the field, and the overall size of the motor can be smaller. On the other hand the strength of the field cannot be varied so one possible option for control is ruled out.

Ferrite magnets have been used for many years. They are relatively cheap and easy to manufacture but their energy product (a measure of their effectiveness as a source of excitation) is poor. Rare earth magnets (e.g. samarium cobalt) provide much higher energy products, and open up the possibility of very high torque/volume ratios. They are used in high-performance servo motors, but are relatively expensive, difficult to manufacture and handle, and the raw material is found in politically sensitive areas. The latest 'super' magnet material is neodymium iron boron, which has the highest energy product and is gradually making an impact on motor design. At present its major handicap is that it can only be operated at temperatures below about 150°, which is not sufficient for some high-performance motors.

Although the magnetic field is essential to the operation of the motor, we should recall that in Chapter 1 we saw that none of the mechanical output power actually comes from the field system. The excitation acts rather like a catalyst in a chemical reaction, making the energy conversion possible but not contributing to the output.

The main (power) circuit consists of a set of identical coils wound in slots on the rotor, and known as the armature. Current is fed into and out of the rotor via carbon brushes which make sliding contact with the 'commutator', which consists of insulated copper segments mounted on a cylindrical former.

The function of the commutator is discussed further below, but it is worth stressing here that all the electrical energy which is to be converted into mechanical output has to be fed into the motor through the brushes and commutator. Given that a high-speed sliding contact is involved, it is not surprising that to ensure trouble-free operation the commutator needs to be kept clean, and the brushes and their associated springs need to be regularly serviced. Brushes wear away, of course, though if properly set they can last for thousands of hours. All being well the brush debris (in the form of graphite particles) will be carried out of harm's way by the ventilating air: any build-up of dust on a greasy commutator is dangerous and can lead to disastrous 'flashover' faults.

The axial length of the commutator depends on the current it has to handle. Small motors usually have one brush on each side of the commutator, so the commutator is quite short, but larger heavy-current motors may well have many brushes mounted on a common arm, each with its own brushbox (in which it is free to slide) and with all the brushes on one arm connected in parallel via their flexible copper leads or 'pigtails'. The length of the commutator can then be comparable with the 'active' length of the armature.

Function of the commutator

Many different winding arrangements are used for d.c. armatures, and it is neither helpful nor necessary for us to delve into the nitty-gritty of winding and commutator design. These are matters which are best left to motor designers and repairers. What we need to do is to focus on what a well-designed commutator-winding actually achieves, and despite the apparent complexity, this can be stated quite simply.

The purpose of the commutator is to ensure that regardless of the position of the rotor, the pattern of current flow in the rotor is always as shown in Figure 3.3.

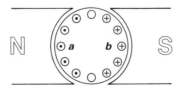

Figure 3.3   *Pattern of rotor (armature) currents in 2-pole d.c. motor*

Current enters the rotor via one brush, flows through all the rotor coils in the directions shown in Figure 3.3, and leaves via the other brush. The first point of contact with the armature is via the segment or segments on which the brush is pressing at the time (the brush is usually wider than a single segment), but since the interconnections between the individual coils are made at each commutator segment, the current actually passes through all the coils via all the commutator segments in its path through the armature.

We can see from Figure 3.3 that all the conductors lying under the N pole carry current in one direction, while all those under the S pole carry current in the opposite direction. All the conductors under the N pole will therefore experience an upward force (which is proportional to the radial flux density B and the armature current I) while all the conductors under the S pole will experience an equal downward force. A torque is thus produced on the rotor, the magnitude of the torque being proportional to the product of the flux density and the current. In practice the flux density will not be completely uniform under the pole, so the force on some of the armature conductors will be greater than on others. However, it is straightforward to show that the total torque developed is given by

$$T = K_T \Phi I \qquad (3.1)$$

where Φ is the total flux produced by the field, and K_T is constant for a given motor. In permanent magnet motors, the flux is sensibly constant, so we see that the motor torque is directly proportional to the armature current. This extremely simple result means that if a motor is required to produce constant torque at all speeds, we simply have to arrange to keep the armature current constant. Standard drive packages usually include provision for doing this, as will be seen later. We can also see from equation 3.1 that the direction of the torque can be reversed by reversing either the armature current (I) or the flux (Φ). We obviously make use of this when we want the motor to run in reverse, and sometimes when we want regenerative braking.

The alert reader might rightly challenge the claim – made above – that the torque will be constant regardless of rotor position. Looking at Figure 3.3, it should be clear that if the rotor turned just a few degrees, one of the five conductors shown as being under the pole will move out into the region where there is no radial flux, before the next one moves under the pole. Instead of five conductors producing force, there will then be only four, so won't the torque be reduced accordingly?

The answer to this question is yes, and it is to limit this unwelcome variation of torque that most motors have many more coils than are shown in Figure 3.3. Smooth torque is of course desirable in most applications in order to avoid vibrations and resonances in the transmission and load, and is essential in machine tool drives where the quality of finish can be marred by uneven cutting if the torque and speed are not steady.

Broadly speaking the higher the number of coils (and commutator segments) the better, because the ideal armature would be one in which the pattern of current on the rotor corresponded to a 'current sheet', rather than a series of discrete packets of current. If the number of coils was infinite, the rotor would look identical at every position, and the torque would therefore be absolutely smooth. Obviously this is not practicable, but it is closely approximated in most d.c. motors. For practical and economic reasons the number of slots is higher in large motors, which may well have a hundred or more coils and hence very little ripple in their output torque.

Operation of the commutator – interpoles

Returning now to the operation of the commutator, and focusing on a particular coil (e.g. the one shown as ab in Figure 3.3) we note that for half a revolution – while side a is under the N pole and side b is under the S pole – the current needs to be positive in side a and negative in side b in order to produce a positive torque. For the other half revolution, while side a is under the S pole and side b is under the N pole, the current must flow in the opposite direction through the coil for it to continue to produce positive torque. This reversal of current takes place in each coil as it passes through the interpolar axis, the coil being 'switched round' by the action of the commutator sliding under the brush.

Each time a coil reaches this position it is said to be undergoing commutation, and the relevant coil in Figure 3.3 has therefore been shown as having no current to indicate that its current is in the process of changing from positive to negative.

The essence of the current-reversal mechanism is revealed by the simplified sketch shown in Figure 3.4. This diagram shows only a single coil, and it should be stressed again that in an actual multi-coil armature, only one coil is reversed at a time, and that the commutator arc is much smaller than that shown in Figure 3.4.

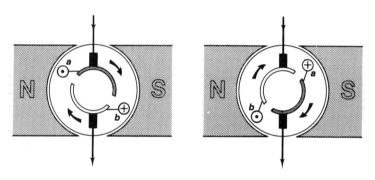

Figure 3.4   *Simplified diagram of single-coil motor to illustrate the current-reversing function of the commutator*

The main difficulty in achieving good commutation arises because of the self inductance of the armature coils, and the associated stored energy. As we have seen earlier, inductive circuits tend to resist change of current, and if the current reversal has not been fully completed by the time the brush slides off the commutator segment in question there will be a spark at the trailing edge of the brush.

In small motors some sparking is considered tolerable, but in medium and large wound-field motors small additional stator poles known as interpoles (or compoles) are provided to improve commutation and hence minimise sparking. These extra poles are located midway between the main field poles, as shown in Figure 3.5. Interpoles are not normally required in

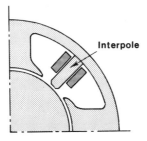

Figure 3.5 *Sketch showing location of interpole and interpole winding. (The main field windings have been omitted for the sake of clarity)*

permanent magnet motors because the absence of stator iron close to the rotor coils results in much lower armature coil inductance.

The purpose of the interpoles is to induce a motional e.m.f. in the coil undergoing commutation, in such a direction as to speed up the desired reversal of current, and thereby prevent sparking. The e.m.f. needed is proportional to the current which has to be commutated, i.e. the armature current, and to the speed of rotation. The correct e.m.f. is therefore achieved by passing the armature current through the coils on the interpoles, thereby making the flux from the interpoles proportional to the armature current. The interpole coils therefore consist of a few turns of thick conductor, connected permanently in series with the armature.

Motional e.m.f.

When the armature is stationary, no motional e.m.f. is induced in it. But when the rotor turns, the armature conductors cut the magnetic flux and an e.m.f. is induced in them.

As far as each individual coil on the armature is concerned, an alternating e.m.f. will be induced in it when the rotor rotates. For the coil ab in Figure 3.3, for example, side a will be moving upward through the flux if the rotation is clockwise, and an e.m.f. directed out of the plane of the paper will be generated. At the same time the 'return' side of the coil (b) will be moving downwards, so the same magnitude of e.m.f. will be generated, but directed into the paper. The resultant e.m.f. in the coil will therefore be twice that in the coil-side, and this e.m.f. will remain constant for almost half a revolution, during which time the coil sides are cutting a constant flux density. For the comparatively short time when the coil is not cutting any flux the e.m.f. will be zero, and then the coil will begin to cut through the flux again, but now each side is under the other pole, so the e.m.f. is in the opposite direction. The resultant e.m.f. waveform in each coil is therefore a rectangular alternating wave, with magnitude and frequency proportional to the speed of rotation.

The coils on the rotor are connected in series, so if we were to look at the e.m.f. across any given pair of diametrically opposite commutator segments, we would see a large alternating e.m.f. (We would have to station ourselves on the rotor to do this, or else make sliding contacts using slip-rings).

The fact that the induced voltage in the rotor is alternating may come as a surprise, since we are talking about a d.c. motor rather than an a.c. one. But any worries we may have should be dispelled when we ask what we will see by way of induced e.m.f. when we 'look in' at the brushes. We shall see that the brushes and commutator effect a remarkable transformation, bringing us back into the reassuring world of d.c.

The first point to note is that the brushes are stationary. This means that although a particular segment under each brush is continually being replaced by its neighbour, the circuit lying between the two brushes always consists of the same number of coils, with the same orientation with respect to the poles. As a result the e.m.f. at the brushes is direct, rather than alternating.

The magnitude of the e.m.f. depends on the position of the brushes around the commutator, but they are invariably placed at the point where they continually 'see' the peak value of the alternating e.m.f. induced in the armature. In effect, the commutator and brushes can be regarded as a mechanical rectifier which converts the alternating e.m.f. in the rotating reference frame to a direct e.m.f. in the stationary reference frame. It is a remarkably clever and effective device, its only real drawback being that it is a mechanical system, and therefore subject to wear.

We saw earlier that to obtain smooth torque it was necessary for there to be a large number of coils and commutator segments, and we find that much the same considerations apply to the smoothness of the generated e.m.f. If there are only a few armature coils the e.m.f. will have a noticeable ripple superimposed on the mean d.c. level. The higher we make the number of coils, the smaller the ripple, and the better the d.c. we produce. The small ripple we inevitably get with a finite number of segments is seldom any problem with motors used in drives, but can give rise to difficulties when a d.c. machine is used as a tachogenerator.

From the discussion of motional e.m.f. in Chapter 1, it follows that the magnitude of the resultant e.m.f. (E) which is generated at the brushes is proportional to the flux (Φ) and the speed (n), and is given by

$$E = K_E \Phi n \qquad (3.2)$$

where K_E is constant for the motor in question.

This equation reminds us of the key role of the flux, in that until we switch on the field no voltage will be generated, no matter how fast the rotor turns. Once the field is energised, the generated voltage is directly proportional to the speed of rotation, so if we reverse the direction of rotation, we will also reverse the polarity of the generated e.m.f. We should also remember that the e.m.f. depends only on the flux and the speed, and is the same regardless of whether the rotation is provided by some external source (i.e. when the machine is being driven as a generator) or when the rotation is produced by the machine itself (i.e. when it is acting as a motor).

When the flux (Φ) is at its full value, equations 3.1 and 3.2 can be written in the form

$$T = k_t I \qquad (3.3)$$

$$E = k_e \omega \qquad (3.4)$$

where k_t is the motor torque constant, k_e is the e.m.f. constant, and ω is the angular speed in rad/sec. The SI units for k_t are

Nm/A, and for k_e the units are volts/rad/sec. (Note, however, that k_e is more often given in volts/1000 rev/min.) In SI units, the torque and e.m.f. constants are equal, i.e. $k_t = k_e = k$. The torque and e.m.f. equations can thus be written

$$T = kI \qquad (3.5)$$

$$E = k\omega \qquad (3.6)$$

Equivalent circuit

The equivalent circuit can now be drawn on the same basis as we used for the primitive machine in Chapter 1, and is shown in Figure 3.6.

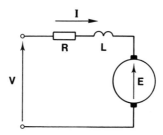

Figure 3.6   *Equivalent circuit of a d.c. motor*

The voltage V is the voltage applied to the armature terminals (i.e. across the brushes), and E is the internally developed motional e.m.f. The resistance and inductance of the complete armature are represented by R and L in Figure 3.6. The sign convention adopted is the usual one when the machine is operating as a motor. Under motoring conditions, the motional e.m.f. E always opposes the applied voltage V, and for this reason it is referred to as 'back e.m.f.' For current to be forced into the motor, V must be greater than E, the voltage equation being given by

$$V = E + IR + L\frac{di}{dt} \qquad (3.7)$$

D.C. MOTOR – STEADY-STATE CHARACTERISTICS

From the user's viewpoint the extent to which speed falls when load is applied, and the variation in speed with applied voltage are usually the first questions which need to be answered in order to assess the suitability of the motor for the job in hand. The information is usually conveyed in the form of the steady-state characteristics, which indicate how the motor behaves when any transient effects (caused for example by a sudden change in the load) have died away and conditions have once again become steady. Steady-state characteristics are usually much easier to predict than transient characteristics, and for the d.c. machine they can all be deduced from the simple equivalent circuit in Figure 3.6.

Under steady conditions, the armature current I is constant and equation 3.7 simplifies to

$$V = E + IR, \quad or \quad I = \frac{(V - E)}{R} \qquad (3.8)$$

We will derive the steady-state torque–speed characteristics for any given armature voltage V, but first we begin by establishing the relationship between the no-load speed and the armature voltage, since this is the foundation on which the speed control philosophy is based.

No-load speed

By 'no load' we mean that the motor is running light, so that the only mechanical resistance is that due to its own friction. In any sensible motor the frictional torque will be small, and only a small driving torque will therefore be needed to keep the motor running. Since motor torque is proportional to current (equation 3.1), the no-load current will also be small. If we assume that the no-load current is in fact zero, the calculation of no-load speed becomes very simple. We note from equation 3.8 that zero current implies that the back e.m.f. is equal to the applied voltage, while equation 3.2 shows that the back e.m.f. is proportional to speed. Hence under true no-load (zero torque) conditions, we obtain

$$V = E = K_E\Phi n, \quad \text{or} \quad n = \frac{V}{K_E\Phi} \tag{3.9}$$

where n is the speed.

At this stage we are concentrating on the steady-state running speeds, but we are bound to wonder how it is that the motor reaches speed from rest. We will return to this when we look at transient behaviour, so for the moment it is sufficient to recall that we came across an equation identical to equation 3.9 when we looked at the primitive motor in Chapter 1. We saw that if there were no load on the shaft, the speed would rise until the back e.m.f. equalled the supply voltage. The same result clearly applies to the real d.c. motor here. We see from equation 3.9 that the no-load speed is directly proportional to armature voltage, and inversely proportional to field flux. For the moment we will continue to consider the case where the flux is constant, and demonstrate by means of an example that the approximations used in arriving at equation 3.9 are justified in practice. Later, we can use the same example to study the torque–speed characteristic.

Performance calculation – example

Consider a 500 V, 10 kW, 20 A motor with an armature resistance of 1 Ω. When supplied at 500 V, the unloaded motor runs at 1040 rev/min, drawing a current of 0.8 A. Note that because this is a real motor, it draws a small current (and therefore produces some torque) even when unloaded. The fact that it needs to produce torque, even though it is not accelerating, is of course attributable to the inevitable friction in the bearings and brushgear.

If we want to estimate the no-load speed at a different armature voltage, say 250 V, we use equation 3.9, giving

no-load speed at 250 V = $(250/500) \times 1040 = 520$ rev/min

Since equation 3.9 is based on the assumption that the no-load current is zero, this result is only approximate. If we insist on

being more precise, we must first calculate the original value of
the back e.m.f., using equation 3.8, which gives

$$E = 500 - 0.8 \times 1 = 499.2 \text{ volts}$$

The corresponding speed is 1040 rev/min, so the e.m.f. constant
must be 499.2/1040 or 480 volts/1000 rev/min. To calculate the
no-load speed for $V = 250$ volts, we must first assume that the
friction torque still corresponds to an armature current of
0.8 A, in which case the back e.m.f. will be given by

$$E = 250 - 0.8 \times 1 = 249.2 \text{ volts}$$

and hence the speed will be given by

no-load speed at 250 V = $(249.2/480) \times 1000 = 519.2$ rev/min

The difference between the approximate and true no-load
speeds is very small, and is unlikely to be significant. Hence we
can safely use equation 3.9 to predict the no-load speed at any
armature voltage, and obtain the set of no-load speeds shown
in Figure 3.7. This diagram illustrates the very simple linear
relationship between the speed of an unloaded d.c. motor and
the armature voltage.

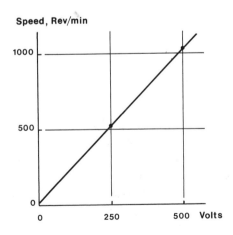

Figure 3.7   *No-load speed of d.c. motor as function of armature
voltage*

Behaviour when loaded

Having seen that the no-load speed of the motor is directly proportional to the armature voltage, we need to explore how the speed will vary if we change the load on the shaft. The usual way we quantify 'load' is to specify the torque needed to drive the load at a particular speed. Some loads, such as a simple drum-type hoist with a constant weight on the hook, require the same torque regardless of speed, but for most loads the torque needed varies with the speed. For a fan, for example, the torque needed varies roughly with the square of the speed. If we know the torque–speed characteristic of the load, and the torque–speed characteristic of the motor, we can find the steady-state speed simply by finding the intersection of the two curves in the torque–speed plane. An example is shown in Figure 3.8.

At point X the torque produced by the motor is exactly equal

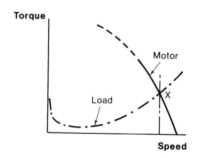

Figure 3.8   *Steady-state torque–speed curves for motor and load showing location (X) of steady-state operating condition*

to the torque needed to keep the load turning, so the motor and load are in equilibrium and the speed remains steady. At all lower speeds, the motor torque is higher than the load torque, so the nett torque will be positive, leading to an acceleration of the motor. As the speed rises towards X the acceleration reduces until the speed stabilises at X. Conversely, at speeds above X the motor's driving torque is less than the braking torque exerted by the load, so the nett torque is negative and the system will decelerate until it reaches equilibrium at X. This